ATMOSPHERIC THINGS

ELEMENTS   *A series edited by Stacy Alaimo and Nicole Starosielski*

ATMOSPHERIC THINGS

ON THE ALLURE OF
ELEMENTAL ENVELOPMENT

DEREK P. McCORMACK

DUKE UNIVERSITY PRESS   Durham and London   2018

Printed and bound by CPI Group (UK) Ltd, Croydon, CR0 4YY
Designed by Amy Ruth Buchanan
Typeset in Chaparral Pro and Knockout by Westchester

Library of Congress Cataloging-in-Publication Data
Names: McCormack, Derek P., author.
Title: Atmospheric things : on the allure of elemental envelopment /
    Derek P. McCormack.
Description: Durham : Duke University Press, 2018. | Series: Elements
    | Includes bibliographical references and index.
Identifiers: LCCN 2017052687 (print) | LCCN 2017061158 (ebook)
ISBN 9780822371731 (ebook)
ISBN 9780822371120 (hardcover : alk. paper)
ISBN 9780822371236 (pbk. : alk. paper)
Subjects: LCSH: Atmosphere. | Balloons—History. | Balloons—
    Scientific applications. | Technological innovations—Social aspects. |
    Technological innovations—Political aspects.
Classification: LCC QC861.3 (ebook) | LCC QC861.3 .M3395 2018 (print) |
    DDC 551.51—dc23
LC record available at https://lccn.loc.gov/2017052687

Cover art: *Marine releases a Combat Sky satellite communication
balloon*, December 3, 2008. Courtesy of Specialist Second Class,
Daniel Barker/U.S. Navy.

For Cillian and Fiachra

According to
the ontological principle
there is nothing which
floats into the world
from nowhere.

—ALFRED NORTH WHITEHEAD,
Process and Reality

I am grateful to the British Academy for the award of a Mid-Career Fellowship which provided me with much of the space and time to complete the bulk of this book. The Oxford University Fell Fund also gave me support at an important juncture. Some of the chapters in this book emerged from encounters with material at archives at the National Center for Atmospheric Research in Boulder, Colorado, the Hoover Institute at Stanford University, the Scott Polar Research Institute at Cambridge, and the Grenna Museum—Andréexpeditionen Polarcenter. I am grateful for the assistance of staff at all of these sites, but especially to Håkan Joriksson in Grenna.

Many people have influenced the emergence of this book, through the exemplary nature of their own work, through words of encouragement, or through invitations to participate in something that got me thinking. There are too many to thank here, but I would like to acknowledge Peter Adey, Ben Anderson, James Ash, Andrew Barry, Jane Bennett, Georgina Born, Tim Choy, William Connolly, Jason Dittmer, JD Dewsbury, Joe Gerlach, Vicky Hunter, Thomas Jellis, Tim Ingold, Caren Kaplan, Alan Latham, Jamie Lorimer, Hayden Lorimer, Damien Masson, Hester Parr, Richard Powell, Emma Roe, Tim Schwanen, Paul Simpson, Kathleen Stewart, Gerard Toal, Nigel Thrift, Phillip Vannini, Sarah Whatmore, and John Wylie. Caren Kaplan's enthusiasm for this project and the eloquent precision of her own work on aerial prospects have been critical in allowing this book to take shape. Audiences at various places, notably those at the STS program at the University of Davis, the Independent Dance Festival in London, the University of Austin Texas, the Telling Stories event at the University of Glasgow, and the Technical University of Braunschweig, also helped bring the ideas in this book to life.

Through the perceptive and generous web making of Sasha Engelmann I have been fortunate to participate in the orbital life of Tomás Saraceno's marvelous work. I am grateful to Tomás, his studio, and his colleagues for generating events and occasions at and through which to think about and with atmospheres and the elemental. I've been inspired in the process by the ideas and writings of Andreas Philippopoulos-Mihalopoulos, Nick Shapiro, Bronislaw Szerszynski, Jol Thomson, and so many others. It has been a particular joy to witness Sasha's remarkable engagement with Tomás's work and her ongoing intellectual journey into questions of the atmosphere and beyond. Thanks also to Jol, Sasha, Tomás, and others for showing me that that chapter 6 is much better when collectively and cosmically enunciated.

Through their close and generous reading, the readers of the manuscript for Duke pushed me to improve the book in important and vital ways: it is a privilege to have such careful attention paid to one's work. Thanks again to Courtney Berger for seeing how a book about balloons did not necessarily need to be a book just about balloons! I am also grateful to Stacey Alaimo and Nicole Starosielski for including me within this wonderful new Elements series. As before, the editorial and production team at Duke, particularly Sandra Korn, have been fantastic.

At Oxford, both the School of Geography and Environment and Mansfield College are full of supportive and collegial people, too many to name here, but they all make a difference. Thanks also to Luke Dickens for doing such a marvelous job when I was trying to finish this book, and to Mark Nethercleft for continuing to ask me if I was going to finish the book about balloons, and again to Gary, Mark, and Steve, even if it is now mostly on WhatsApp.

Early versions of some of the ideas in this book were piloted in the pages of the following journals: *The Annals of the Association of American Geographers*; *Cultural Geographies*; *Environment and Planning D: Society and Space*; *Performance Research*; and *Transactions of the Institute of British Geographers*; and in chapters in the collections *Nonrepresentational Methodologies: Re-Envisioning Research* (Routledge, 2015), edited by Phillip Vannini, and *Moving Sites: Investigating Site-Specific Dance Performance* (Routledge, 2015), edited by Victoria Hunter. I am grateful to the reviewers, readers, and editors of all of these publications for helping me to develop ideas.

Because I don't say it often enough: thanks to my parents, and to John, Claire, Mari, and Aidan. To Cillian, Fiachra, and Andrea for helping me remain variously tethered and buoyant.

UP

At the end of his speech, John Kerry stepped away from the lectern, and looked up. At that point, 100,000 red, white, and blue balloons were scheduled to drop on Kerry and the delegates attending the 2004 Democratic National Convention in Boston. The balloons seemed hesitant, however, reluctant to fall from the roof of the convention center. The interval between the end of the address and the balloon drop lengthened, and convention producer, Don Mischer, a member of the Event Industry Hall of Fame, recipient of fifteen Emmy awards, offered some encouragement: "Go balloons! Go balloons! Go balloons! I don't see anything happening. . . . Go balloons! Go balloons! Go balloons! Stand by, confetti. Keep coming, balloons. More balloons. Bring it—balloons, balloons, balloons! We want balloons, tons of them. Bring them down. Let them down!"[1]

Some balloons did fall, but only in small clusters, or as scattered individuals. Without their arrival in any significant numbers, the affective intensity of the moment immediately following Kerry's address began to dissipate, and Mischer could feel it: "We need more balloons. I want all balloons to go, God damn it! Go confetti. More confetti. I want more balloons. What's happening to the balloons? We need more balloons. We need all of them coming down. Go balloons! Balloons! What's happening, balloons? There's not enough coming down! All balloons! What the hell? There's nothing falling! What the fuck are you guys doing up there? We want more balloons coming down!"

At some point, someone noticed that Mischer's words were being broadcast live on CNN. The feed was cut, followed quickly by an apology and explanation by the network. It might be a bad omen, the political commentators said. They recalled that before he lost to Reagan, Jimmy Carter also had a bad time with balloons: Carter's speech to the 1980 Democratic Convention was followed by an underwhelming balloon drop, and unlike

FIGURE I.1 Hillary Clinton and Tim Kaine on stage as the balloons drop at the DNC in Philadelphia, Pennsylvania, July 28, 2016. Photo by Lorie Shaull. Wikimedia Commons.

Reagan, he received no convention bounce. Reagan, in contrast, seemed to know how to handle balloons. He, and the things with which he surrounded himself, felt at home at a party: Republicans had initiated the tradition of the balloon drop at Eisenhower's 1956 convention.[2] With the exception of a failed balloon drop in 1964, the Democratic Party began using the technique only in 1980, and then not very successfully.

The comparative lack of balloons left the moment at the end of Kerry's speech feeling a little empty, feeling like it lacked something. Convention centers, like airship hangers, are big empty spaces of enormous physical volume. Ticker tape, confetti, and balloons allow the volume of these spaces to become atmospheric in distinctive ways. But balloons seem particularly alluring and captivating in this respect, as the more recent 2016 DNC demonstrated (see figure I.1.). Such experiences are highly choreographed. As Treb Heining, organizer of some of these balloon drops, puts it, "It's like a symphony—you've got to have a system that works. It's a celebratory thing, it's the final thing people see, and it's something everyone anticipates. I gotta believe there is a connection between the balloon drop and the convention bounce."[3] While it is unlikely that Kerry's underwhelming balloon drop made a crucial difference to the eventual failure of his candidacy, it contributed to an atmosphere of affective uncertainty around its prospects, to the feeling of unease about his affective capacities. If his balloons were not decisive enough to fall, if he could not control such simple things, if he was incapable of feeling the simple but shared force of their allure, then what did that say about him and his capacity to lead the nation?[4]

DOWN

The line of white illuminated balloons was approximately nine miles long, snaking through Berlin. It formed *Lichtgrenze*, or *Border of Light*, designed by artist Christopher Brauder and his brother, filmmaker Marc Brauder. The installation was at the heart of a series of planned events marking the twenty-fifth anniversary of the fall of the Berlin Wall in 1989. Filled with helium, the eight thousand biodegradable balloons were positioned on 3.6-meter-high poles, matching the approximate height of the wall.

For a few days before the anniversary, this line of balloons became a feature in Berlin, redrawing the line of a structure whose physical and affective presence had in many ways faded and in some respects been forgotten, save for a discreet line of bricks embedded in the ground along the line of the wall. In anticipation of the anniversary, and lured, perhaps, by the presence of the balloons, people walked, cycled, and jogged along the line, (re)familiarizing themselves with the arbitrary, angular geography of its bisection of the city. At the appointed time on the evening of November 9 the balloons were released into the sky, one by one, to symbolize the breaching of the wall. As they floated up, they trailed short lines to which messages written by volunteers had been attached. The balloons and the messages floated up into the night sky, disappearing, withdrawing from the crowd below, with the question of whether they would ever be found again remaining an open one. In doing so they performed the dream of a distinctive form of memorialization, in which an event is marked in the feeling of an atmosphere that never materializes as an entity.[5]

FLYING AROUND

In mid-2013 Google announced it had been undertaking experiments with balloons launched from a site in Christchurch, New Zealand. The launches took place under the aegis of Project Loon. Conceived by Google as a "network of balloons travelling on the edge of space,"[6] Project Loon is designed to provide a cost-effective and feasible technological solution to the problem of how to bring the Internet to the two-thirds of the world's population not currently connected. Specially designed antennas on the ground communicate with the balloons, which in turn talk to each other and to the base station of the local Internet provider. Floating in the stratosphere, the balloons are steered from Loon "mission control" in California by taking advantage of the winds moving at different directions and altitudes. Initially thirty balloons were launched over New Zealand, and a small group of "pilot testers" were recruited on the ground to assess the reliability and speed of the Internet coverage they provided.

The engineers behind the project were enthusiastic about its possibilities, and about the distinctive mode of collaboration it promised. While people and devices were involved, so also were the elemental conditions of the atmosphere. As one of the project team members, the wonderfully aptly named Astro Teller, put it, "We can sail with the wind, and shape the waves and patterns of these balloons, so that when one balloon leaves, another balloon is set to take its place."[7] For the Loon team, this was an experiment with a new kind of stratospheric infrastructure. An experiment, currently ongoing, with the technical capacity to choreograph the position and possibilities of a relatively simple device—the balloon—in the midst of the elemental variations of the atmospheres in which forms of life on Earth are enveloped.

THIS BOOK IS ABOUT the relation between atmospheres and envelopment. It uses a deceptively simple thing—the balloon—as a speculative device for exploring how atmospheres are disclosed, made palpable, and modified through practices and experiences of envelopment. By atmospheres I mean elemental spacetimes that are simultaneously affective and meteorological, whose force and variation can be felt, sometimes only barely, in bodies of different kinds. By "envelopment" I mean two related things. The first is a condition. From the point of view of a body (which could be but need not be human), envelopment is the condition of being immersed within an atmosphere. Being enveloped is a condition that can be sensed, although it is not always. Such sensing is always partial, insofar as an atmosphere is never fully disclosed to something immersed in that atmosphere—hence its allure. Nor is this sensing a distinctively human capacity: it is found in bodies, entities, and agencies of myriad kinds. For instance, variations in atmospheric humidity are sensed and expressed in the wrinkling shape of paper as much as they are sensed and expressed through the clamminess of skin.

Envelopment is not only a condition of being immersed in an atmospheric milieu, however. To think of it solely in these terms is to risk making too clear-cut and static a distinction between entities and atmospheres, with the former floating in or being surrounded by the latter. Envelopment therefore also names a process: a kind of "extrusive" shaping of things in relation to an atmospheric milieu.[8] Because of this, in this book I also use "envelopment" to denote a process of fabrication through which the folding of a membrane of some kind generates something within an atmospheric milieu with the capacity to sense variations in that milieu. My argument is that envelopment is critical for thinking about atmospheres because it allows us to hold in generative tension a relation of material continuity between entities and the elemental conditions in which these entities are immersed and in which they participate. Envelopment is a process for sensing a condition; it is a process through which atmospheric things emerge whose form, shape, and duration depends upon their capacity to sense and respond to the atmospheres in which they are immersed. It is the process by which entities emerge within a milieu from which they differ without becoming discontinuous, in the same way that a cloud is a process of differentiation within an atmosphere without necessarily being discontinuous with it.

This emphasis on envelopment involves a certain kind of formalism, one inspired to some degree by the work of Peter Sloterdijk.[9] However, this is a formalism that is always in formation. Inspired just as much by Michel Serres and Luce Irigaray, it is about following shapes of change as much as about attending to enduring entities.[10] As these thinkers remind us in different ways, forms of life and bodily capacities can be defined by their relative envelopment in relation to the elemental conditions of air. To foreground this is not to emphasize building rather than dwelling; rather, it is to focus on processes of fabricated envelopment as a necessary dimension of both, as part of attention to the practices via which life becomes air-conditioned from inside and out.[11]

Envelopment is therefore not only the condition of being within atmospheres; it is also, as James Ash reminds us, a process through which atmospheres are disclosed and become palpable as elemental conditions of experience via different configurations of bodies, materials, and devices.[12] This is simultaneously a technical, aesthetic, and ethicopolitical matter of concern. It is technical because it involves fabricating envelopes that modify and mediate the exposure of bodies to an outside.[13] It is aesthetic because in certain circumstances it exposes bodies to elemental forces that, in remaining withdrawn from apprehension or cognition, are characterized by a degree of allure or enchantment.[14] And it is ethicopolitical insofar as the question of how bodies can be exposed to elemental forces is often circumscribed by assumptions about what kinds of bodies can be exposed to these forces, and under which circumstances.

ENVELOPMENT PROVIDES AN IMPORTANT way of speculating, experimenting with, and disclosing atmospheres. For many of us in the social sciences and humanities, atmosphere has become one of the most theoretically and empirically alluring of concepts.[15] It provides a way of grasping affective spacetimes that acknowledges their force and palpability even if they remain vague and unformed.[16] It connects (or provides what Tim Ingold calls the "denominator" between) the affective as a field of potentially sensed palpability with the meteorological as the variation in the gaseous medium in which much life on Earth is immersed.[17] It resists any reduction, either to the terms of individual experience, or to the status of an object. Rather, and residing as it does in excess of and between bodies as much as within them, thinking about and with atmospheres requires us to develop a conceptual vocabulary for distinctive kinds of materialist accounts of spacetimes, while also drawing our attention to how these spacetimes become the focus of intervention, action, and experiment.

In this book I pursue the development of such a vocabulary by exploring different possibilities for apprehending the properties and qualities of atmospheres through envelopment. This requires engagement with a number of important questions precipitated by attention to atmospheres. Some of these questions are ontological: they concern the different ways in which the speculative realities of atmospheres can be grasped. In *Atmospheric Things* I work between two currents of thinking about atmospheres. The first is an atmospheric materialism geared toward exploring the qualities and forces of the diffuse, airy, affective spacetimes that operate across, between, and beyond bodies and things.[18] And the other is an entity-oriented ontology that foregrounds the qualities and properties of nonhuman things or objects as the elemental basis for a philosophical, ethical, and political account of reality in which the human is no longer placed at center stage.[19] *Atmospheric Things* is positioned between both: it develops an account of agencies excessive of the category of entity—the atmospheric—through attending to what Jane Bennett calls the "force of things."[20] As such, it resists the logic of "entification" that underpins some object- or entity-oriented accounts of reality (or what Tim Ingold calls a "blobular ontology") while nevertheless attending to the processual emergence of entities with capacities to act and sense the atmospheric.[21] Rather than trying to figure out what kind of thing or entity an atmosphere is, or identifying and cataloguing atmospheres as if they were relatively stable envelopes of experience, my concern here is with accounting for the potentially palpable affective materiality of the elemental spacetimes in which bodies—human and nonhuman—are enveloped, spacetimes that never present themselves as fully tangible, discrete, or unified entities.

If atmospheres raise ontological questions, they also pose empirical and methodological ones. At stake here is the question of how the force of the atmospheric can be "grasped" through the variations, perturbations, or affects it generates in entities of any kind without necessarily reducing this process to the terms of a representational or cognitive relation. To attend to and through the atmospheric requires, as Kathleen Stewart reminds us, the cultivation of particular modes of attunement.[22] As Timothy Choy and Jerry Zee argue, following Stewart, this is about a "form of attention that is also a mode of relation, a way of being suspended. This form of thought looks up and around, at plumes, clouds, and sky. It looks inward through the vital interiors that render bodies channels, containers, and filters for airs and the things they hold. More significant than the directionality of its gaze, however, is its manner of attunement to the potentials of substances to shift from states of settlement or condensation to ones of airborne agitation, to settle again in time, or to activate a reaction, somewhere else."[23]

On one level, then, atmosphere appeals to an expanded empiricism through which capacities to sense vague and fleeting variations in elemental conditions become foregrounded. In turn, this raises questions about what it means to sense, about what is being sensed, and about how envelopes of sensing can be stretched through forms of experiment. It requires us to think about how far sensing might go beyond the sphere of human capacities and experience, and about the techniques, technologies, and devices that facilitate such forms of sensing, while also generating opportunities for making something palpable in experience.

In thinking through these questions there is a temptation to invoke atmosphere as shorthand for an empiricism that privileges presence, immediacy, and immersion.[24] Atmosphere can be too easily affirmed as a concept for reclaiming some kind of authentic experience of a world as a counter to the alienating and distancing tendencies both of contemporary life and of certain flavors of critical thinking. For instance, Hans Ulrich Gumbrecht writes, "Concentrating on atmospheres and moods offers literary studies a possibility for reclaiming vitality and aesthetic immediacy that have, for the most part, gone missing."[25] While experiences of vitality are possible, atmosphere complicates claims about what Mitch Rose calls "dreams of presence."[26] Even as it seems to privilege immersion, atmosphere is haunted by something in excess of immediacy: that which is withdrawn from apprehension, even when actively manipulated by devices and technologies that hold out the promise of immersion.[27] Moreover, the question of how atmospheres are sensed is complicated further because this question can also be extended to nonhuman entities and agencies, from animals to technological devices, with the capacities to sense processes, gradients, and variations. Atmospheres do not only register as the "intersubjective" feeling of a variation in an elemental milieu; they also register through what might be understood as an "interobjective" capacity to sense such variation.

These ontological and methodological questions are also entangled in political questions. Some of these have to do with how different kinds of bodies are differentially enveloped by atmospheres. And some have to do with how the capacities to generate, sense, and modify atmospheres are distributed. At stake politically in accounts of atmosphere are the terms of the relations between different bodies, the infrastructures and devices that condition the atmospheres in which they move, and the capacities of these bodies to exercise some influence over these conditions. Atmospheres are necessary conditions for the security of forms of life but can also be threatened by those forms of life.[28] Because of this, atmospheric politics not only involves mobilizing capacities to generate senses of involvement around the promise of being and feeling enveloped; as scholars such as Ben Anderson, Peter Adey, Marjin Nieuwenhuis, and Andreas Philippopoulos-Mihalopoulos have argued, it also involves capacities to withdraw from, to puncture, and to modify exposure to atmospheres that can just as easily function to numb, distract, and occlude as they can to disclose, make palpable, or heighten attunement.[29] Atmospheric politics operate through experiments (some more careful than others) with different techniques or devices for heightening or diminishing capacities to modify the conditions of envelopment that sustain diverse forms of life. To foreground envelopment is therefore by no means to naively affirm its capacity to generate shared atmospheric spacetimes of affective solidarity. Processes of envelopment are differently implicated in an infrastructural politics of immersion, awareness, and exposure that draws some bodies in and excludes others.

Ethically, at stake here is the question of how to live in relation to the elemental energies of air and atmosphere, and in ways that balance the twin requirements of envelopment and exposure as necessary conditions for the flourishing of forms of life.[30] This question is not only a matter of theoretical formulation or distanced judgment. It also involves pursuing forms of modest experiment that make atmospheres explicit in different ways while also distributing and stretching capacities to sense their force. As scholars such as Timothy Choy and Sasha Engelmann have suggested, this might involve the elaboration of a "poetics of air" through which variations in the force of atmospheres and the capacities of bodies can unfold together.[31] It might involve cultivating capacities to sense variations in the affective-meteorological spacetimes in which forms of life take shape, not necessarily in order to generate spaces of common ground but in order to produce infrastructures, composed of arrangements of devices, concepts, and bodies; these are infrastructures for generating value from an elemental commons.[32]

In this book I pursue such possibilities by tracking experiments with envelopment that open up different modes of being and becoming more attuned to the elemental. I invoke the elemental in order to signal the entanglement of a number of matters of concern. The elemental gathers together concerns about the meteorological and affective force of atmospheres as environmental milieus; ontological concerns about the nature of things; as well as concerns about the properties and capacities of particular substances and materials, including gases. Envelopment is elemental, in this sense, because it is a condition and process that mixes these different matters of concern. While experiments with elemental envelopment are by no means always benign, animating this book is the possibility that under the right circumstances, such experiments might renew a certain kind of atmospheric awareness of the elemental conditions that mix the affective, the meteorological, and the technical. In the process, such experiments might generate resources for reimagining, reinventing, and fabricating collective forms of atmospheric life.

MY METHOD FOR EXPERIMENTING with elemental envelopment is to focus on the shape of a particular thing, at once familiar and uncanny, modest and sometimes spectacular—the balloon. I mobilize the balloon to show how the process of envelopment affords important opportunities for experimenting with the condition of being enveloped by elemental atmospheres. Put another way, the balloon figures here as a speculative device for doing atmospheric things. By "atmospheric things" I mean objects, processes, or events that in some ways disclose, generate, or intensify the condition of being enveloped by the elemental force of atmospheres. Critically, by developing the concept of atmospheric things here I designate a sense of something happening as much as I do an entity or object.[33]

The balloon is not the only device for doing atmospheric things. But it is a particularly useful and alluring one, as the opening vignettes reveal. The phenomenon of the balloon drop is an obvious reminder that the choreographed staging of atmospheres is central to the generation of a sense of occasion that gathers around the performance and promise of an event. While the importance of this promise is heightened by the political affects of a presidential convention, it also gathers around innumerable other events, many of which are more familiar, from product launches to birthday parties. In the case of each event, because we expect some sense of an atmosphere, we feel its absence as a failure and a sense of deflation, as both John Kerry and Don Mischer did. Of course, such events, and the atmospheres that gather around them, do not need to be light, happy, or positive; they can be heavy, sad, or disturbing. However different, the allure of atmospheric things resides in a distributed and vague spacetime of palpability: it resides in the sense that something unformed is happening in the elemental milieu in which bodies move.[34]

While the balloon may participate in the generation of the palpable sense of an atmosphere, it also has an uncanny capacity to point to the limits of our envelopes of atmospheric experience. It holds in tension the capacity to generate a condition of palpable envelopment that itself remains somewhat withdrawn from the sensory grasp of the bodies and entities it envelops. It reminds us that atmospheric things are therefore alluring precisely because something of them always remains beyond cognition or tangibility. Something of them always continues to be evanescent and unsubstantiated; they are always partially vague and vaporous, forms variously precipitating and dissolving, becoming present while also withdrawing from bodies. As the balloon event marking the twenty-fifth anniversary of the fall of the Berlin wall suggests, an act of release can become, through the disappearance of an object into the sky, an event through which an atmosphere becomes palpable through a certain kind of withdrawal.

Through this act of release an atmosphere becomes palpable without becoming permanent or fully present. In the air above Berlin, the release and withdrawal of a series of discrete devices generates something atmospheric that never substantiates itself in the form of an entity. Atmospheric things, in other words, remain excessive of efforts to both grasp them in thought and to define their material form even when they are disclosed, generated, or modified by particular arrangements of objects.

It would be wrong to suggest that atmospheres are important only in relation to the affective or sensory capacities of humans, however.[35] Or to think of them as purely emotional phenomena. Atmospheres are also meteorological, and not only metaphorically so: atmospheres are elemental spacetimes from which forms (such as clouds) and patterns (such as winds) emerge and into which they dissolve. They are elemental spacetimes within which myriad things, living and nonliving, are immersed and enveloped. Some, but not all, of these forms and patterns are of course palpable in the mundane worlds of living things, including humans. But some are sensed only remotely, via devices operating beyond the orbit of everyday experience. This means that experiments with envelopment in the form of the balloon—as a device for doing atmospheric things—are alluring in part because they reveal the properties and qualities of the *meteorological* atmosphere. The balloon has been and continues to be central to how this atmosphere has been disclosed and understood. Google's Loon experiment is a further iteration of this: it is based upon the dream of disclosing the dynamics of the stratosphere via an operational assemblage of devices that, in turn, also stretch the kinds of experiences that cohere around screens and surfaces of different kinds as part of contemporary forms of atmospheric media.[36]

But even as these experiments are framed and financed in technical and commercial terms, the engineers working on Project Loon describe what they are doing in terms that are far from the narrow, objective, metric understanding of atmosphere by scientists and engineers:[37] theirs is a joyous, emphatic, exuberant sense of the meteorological atmosphere as an elemental medium for doing things. They express a way of working *with*, and articulate a certain feeling *for*, the elemental conditions of the atmosphere as collaborators in distinctive kinds of speculative experiment. For the Google engineers, the allure of experimenting with atmospheric things is therefore about configuring different combinations of practices, devices, and technologies in order to generate capacities to sense, feel, and value elemental spacetimes.

I dally with this speculative promise here but do so in order to pursue possibilities for generating other sources of value. In the process I deploy the balloon as a philosophical as much as a technical device. Philosophically, the balloon is a device for speculating and moving between the primacy of intensive process and the extrusive presence of affective materials taking shape but never quite stabilizing as objects. Thinking with the balloon allows me to settle, at times, on the form of a thing as a lure for thinking. There is some overlap here with what Ian Bogost calls "alien phenomenology," in as much as my approach is to deliberately foreground the strange and uncanny sense of an everyday thing, recognizing that something of the balloon is always withdrawn from or inaccessible to other entities, while also working with its capacities to move us to think.[38] And yet, while the form of the balloon provides a constraint for thinking, my use of it here is not quite the same as what others have called "following the thing."[39] Rather, I attend to the shape of this thing as a method for remaining attentive to how a process—envelopment—takes form in ways that can sensitize us to the elemental force and allure of atmospheric things whose condition exceeds the category of entity. Put another way, the form of an object becomes a way of holding in view, and becoming attuned in certain circumstances, to what Kathleen Stewart calls the "unformed objects" of atmospheres.[40]

My argument in *Atmospheric Things* is informed by engagements with and experiences of experiments with the balloon as a device for doing atmospheric things. I draw upon archives of visual and textual material related to particular episodes of balloon travel and experiment. This is layered, in turn, by readings of depictions of the balloon in various cultural forms and genres, including literature, poetry, and cinema. At the same time, and building upon my approach in an earlier book, the arguments here are informed by minor experiential encounters at sites associated with different scientific, expeditionary, and aesthetic experiments.[41] *Atmospheric Things* does not so much offer a cultural history of the balloon or ballooning, however, not least because it remains focused on a largely western, modern context, and is confined to a period beginning in the late eighteenth century with the first practical balloon flights in Europe.[42] The rich if contested traditions of doing atmospheric things with sky lanterns in China, Taiwan, Thailand, and Brazil are for the most part absent here.

Nevertheless, these qualifications notwithstanding, throughout the book, the balloon performs as a speculative device with which to sound out how envelopment makes the elemental force of atmospheres potentially present and palpable (although never fully). As a device for thinking-with, the balloon has some obvious risks, however. Its apparent levity might appear too trivial or whimsical to bear the weight of the work demanded of it. There is the danger, also, that it becomes a vehicle for a speculative philosophy based upon ascension, transcendence, and eleva-tion. And, usually lacking dirigibility, it has a waywardness that short-circuits directness of thought, or clouds the development of a clear line of argument. Acknowledging these risks, in *Atmospheric Things* I work with them. I show how, rather than solely an object of levity and lightness, the balloon can also become a device for modulating and distributing geogra-phies and experiences of grief, sadness, and terror. I show that its capacity to facilitate ascension is not reducible to the terms of a critique of the god trick as a mode of thinking and seeing that privileges distance and objectivity;[43] it also facilitates practices for sensing envelopment in ways that complicate such well-rehearsed critical moves, although in perhaps surprising ways. Critically, I show how a lack of dirigibility is not necessar-ily a disadvantage: the balloon moves with the conditions in which it is immersed, offering an image of thinking and moving sustained by the currents and trajectories of the atmosphere.

As the philosopher Michel Serres reminds us in a reading of the novels of Jules Verne, the balloon offers a vehicle for a style of semirandom wan-dering that links together different spacetimes according to the quality of aerial/earthly circumstances, albeit within certain limits. These are jour-neys that become a series of landing sites strung together by lines of open-ended drift. These are journeys whose arcs can be traced via the ongoing processes of ascending and descending, expansion and contraction. In this sense, the narrative spaces of which the balloon as a technology of move-ment and mobility are generative—as a space*temps* machine—are not strictly directional. They are composed of all kinds of loops and deflections, shortcuts and wormholes, creating opportunities for telling stories that fold in and back on themselves in multiple ways. These are what Serres, particularly attentive to the qualities of Verne's spacetime machines, calls "strange journeys." As Serres reminds us, to travel with the balloon is to undertake "voyages through a plurality of spaces, by means of an exfoli-ated multiplicity of maps." To tell the story of these journeys, "one must lose oneself from space to space, from circle to circle, from map to map, from world-map to world-map."[44] Serres's own writing exemplifies the qualities of these journeys, with the topological logic of his pages and paragraphs, frequently tracking back and forth across times and places.

As Laura Salisbury puts it, this style reflects Serres's own view that the "philosopher is simply attentive to the way in which things become unexpectedly close or distant within a temporality that is chaotic and turbulent, a time that is more meteorological in its movements than classically historicist."[45] So, if we take Serres at his word, the balloon provides an imaginative-conceptual vehicle via which to undertake journeys—and perform histories and geographies—that move across topological surfaces stretching between circumstances that gather as—and around—atmospheric things.

The balloon is a partially dirigible device for experimenting with the relation between envelopment and the elemental force of atmospheres. By "experiment" here I mean, broadly, a loosely organized set of practices—which include thinking—geared toward the undertaking of an operation whose outcome, while potentially predictable, remains unguaranteed. I take experiment to be an ethos as much as a well-policed set of technical protocols. Equally, experiments are always circumstance-specific, taking shape as localized arrangements of bodies, devices, and concepts. Experiment is therefore implicated in different ways in how I approach the balloon as a device for doing atmospheric things. On one level, the balloon has and continues to be the locus of scientific experiments with disclosing the condition of the atmosphere. But it has also been used, and continues to be used, in artistic experiments. Admittedly, some of these experiments are reminders that envelopment is by no means always generative. Notwithstanding this, in *Atmospheric Things* I also want to sound an affirmative tone. I follow Kathleen Stewart in affirming that "things hanging in the air are worth describing" because in doing so it might become possible to stretch the envelope of ways of sensing and inhabiting atmospheric worlds.[46] Through tracking its deployment in a range of practices in different domains of expertise, my aim is to show how, as a device for doing atmospheric things, the balloon is implicated in experiments that disclose the aesthetic, ethical, and political relations between different forms of life and the elemental envelopes that sustain them.

MY ARGUMENT IN *Atmospheric Things* begins, in chapter 1, with a discussion of the importance of envelopment as a condition and a process for thinking about atmospheres. My claim is that focusing on envelopment provides a way of thinking between two broad trajectories of thinking: namely, an atmospheric materialism and an entity-centered ontology. In this context, envelopment provides a way of holding on to the processual fabrication of the force of atmospheric things while avoiding the reduction of atmospheres to the category of entity or object.[47] This claim provides the basis for chapter 2, in which I show how, in the shape of the balloon, envelopment affords distinctive modes of sensing the elemental force of an atmospheric milieu whose variations become palpable in bodies as a disquieting kind of stillness in motion. However, rather than affirming pure immersion, thinking about this sensing involves reckoning with a degree of remoteness. In chapter 3 I develop this claim further by focusing more specifically on allure as it relates to atmospheric things. My claim here is that while important, it is not enough to argue that the allure of something resides in the fact of its withdrawal from the world.[48] Understanding the allure of elemental envelopment also requires attention to the specific ways in which it is fabricated. The allure of elemental envelopment is therefore not only a metaphysical problem; it is also a technical one requiring attention to the capacities of specific materials and practices of envelopment to render bodies susceptible in different ways to a force that exceeds their grasp.

Understanding atmospheric things also requires attention to the distinctive kinds of acts around which they gather. In relation to the balloon, one of the most important of these acts is release. In chapter 4 I use the act of release as a way of thinking about how the allure of elemental envelopment resides in the tension between holding on and letting go. To do this, I turn a nominally happy object into a surprisingly sad one by considering acts of balloon release associated with occasions of grief and loss. These occasions allow the sensing of the atmospheric spacing of love as a condition both of the envelopment of being and of being exposed.

In chapters 5 and 6 my attention turns to thinking about how to account for the spatiotemporality of atmospheres in ways that do not reduce it to the shape of a three-dimensional entity. I develop this in chapter 5 by focusing on the question of volume, arguing that we can understand the extent and intensity of atmospheres by developing a differentiated sense of volume and, more specifically, by distinguishing between the volumetric and the voluminous. In chapter 6 I elaborate upon the sonorous associations of volume further by exploring sounding as a technique for rendering atmospheres explicit, and for experimenting with the limits of the atmospheric, not as that which can be sensed but as the very threshold of sense-ability.

Chapters 7, 8, and 9 explore, albeit in different ways, how the atmospheric becomes political and ethical through forms and practices of envelopment. In chapter 7, I explore how the tensions between envelopment and exposure, ground and air, are translated into structures and bodies. Doing so allows me to explore a minor archive of working with the condition and process of envelopment as it is differentially experienced. It also allows me to point to the value of experiments that unsettle and unground these experiences. In chapter 8 my attention turns to the question of how the meteorological atmosphere becomes a medium for the distribution and dispersal of objects, ideas, and affects designed to target and generate the affective and technical infrastructures that sustain different forms of life. In doing so I argue that the politics of envelopment is not only a matter of elevated vision or persistent presence; it also involves experimenting with the extra-territorial logics of drift as an alternative atmospheric mode of address. In chapter 9, I delineate further the kinds of atmospheric politics that gather around experiments with envelopment by focusing on questions of the elemental. I do so by exploring recent and ongoing experiments with stratospheric infrastructures and devices that disclose and sense elemental variations in atmospheres, mobilizing the figure of the angel as part of my efforts to make sense of these experiments. Even if they raise difficult questions about the question of the atmosphere as a form of commons in motion, these experiments also draw our attention to the question of how, and to what ends, we might generate value from experiments with elemental envelopment.

1

What happens when something appears suddenly in the air, something never seen before? Donald Barthelme's short story "The Balloon" describes the sudden appearance of a vast inflated envelope over Manhattan which expands until it covers much of that island.[1] The presence of this balloon, initiated and controlled by an unnamed, unidentified, and omniscient figure with a number of engineers in his employment, provokes many responses. As the event of its appearance unfolds above a bemused and perplexed city, there is little agreement about the real significance or purpose of the balloon; before long, any search for the true meaning of the situation in which the city now finds itself subsides. The narrator even suggests that "situation" does not describe the event of the balloon's appearance because this term already implies "sets of circumstances leading to some resolution, some escape of tension."[2] For the narrator there is no such situation, no play of forces to be untangled, "simply the balloon hanging there."[3] Consequently, for those for whom it becomes a matter of concern, conjecture, or involvement, "it was agreed that since the meaning of the balloon could never be known absolutely, extended discussion was pointless."[4]

No matter how vexing the "apparent purposelessness of balloon" may have been, its presence is not met with a generalized state of indifference;[5] what emerges instead are different ways of relating, affectively, to this enormous aerial thing and to the atmospheres it generates across the city. The failure to find meaning in the event of its appearance is met also with the affirmation of the matter-of-factness of the object-ness of the balloon as it gathers and generates a cloud of affective relations. But rather than a static entity, in the sky above Manhattan the balloon becomes something of a shape-shifter, one whose very form seems to evade delineation. As the narrator recounts, "It was suggested that what was admired about the balloon was finally this: that it was not limited or defined. Sometimes a bulge, blister, or sub-section would carry all the way east to the river on its own initiative, in the manner of an army's movements on a map, as seen in a headquarters remote from the fighting. Then, that part would be, as it were, thrown back again, or would withdraw into new dispositions."[6]

Barthelme's balloon story has been described as a postmodern fable of sorts, sending up as it does the dream of reaching any kind of meaningful narrative resolution.[7] It can be deployed rather differently however: as an imaginative lure for thinking about the relation between atmospheres—as diffuse, unformed spacetimes—and the entities or bodies enveloped by those atmospheres. In the air above Manhattan, Barthelme's balloon is a discrete presence, a "concrete particular, hanging there." It is also an event around which an atmosphere gathers. And yet, as the story unfolds, the balloon becomes a dynamic set of undulating movements and directional tendencies, a shape-shifter composed of both extensive relations and processes of internal differentiation that are neither reducible to an entity nor so diffuse as to be unfelt. Barthelme's balloon, in other words, becomes a device for doing atmospheric things. As a lure for thinking, it becomes a strange attractor: it pulls us toward the possibility of thinking about envelopment as a process through which shapes of change emerge in the tensed space between entities and their atmospheric excess.

THINGS BECOMING ATMOSPHERIC

Despite some expressions of hostility toward the balloon, its presence in the air above Manhattan begins to elicit the emergence of a form of "public warmth." Indeed, the force of this warmth proves strong enough to dissuade the authorities from removing or destroying the balloon.[8] Equally, a certain distribution of feeling emerges across and between the balloon and the people on the ground, a distribution that itself becomes a matter of debate and speculation: "It was argued that what was important was what you felt when you stood under the balloon; some people claimed that they felt sheltered, warmed, as never before, while enemies of the balloon felt, or reported feeling constrained, a 'heavy' feeling."[9] In short, in the air above Manhattan, Barthelme's balloon gathers around itself something that might be grasped as an atmosphere, a vague, differentiated, yet definitely palpable affective spacetime.

The appeal of atmosphere as a concept for naming such spacetimes is almost intuitively obvious, invoking as it does the vague sense of a distributed envelope of feeling, mood, or emotion; how else, indeed, do we account for the diffuse condition of potential palpability that seems to reside between bodies and things?[10] As a way of naming affective spacetimes, "atmosphere" has an ordinary familiarity and everyday currency that precedes its growing visibility as a speculative concept. Its importance is already felt, understood, and acted upon in myriad contexts; anyone who has ever been at or organized any kind of "event," or "occasion," or "do," realizes this. Equally, many practices and crafts, from architecture to advertising, party planning to propaganda, are organized around the promise of acting upon, generating, and staging atmospheres.[11] Their goal is not simply to provide a kind of background against which events happen. The twin promise of atmospheric envelopment is to move bodies to become more or less responsive to their conditions, and to modulate their capacities to act into and within these conditions.[12]

Such is the familiarity of the terms, and the frequency with which they are deployed, that it might seem perplexing to anyone outside the social sciences and humanities that "atmosphere" and "the atmospheric" have recently been (re)discovered, or considered worthy of renewed attention. It might also seem odd that to claim that the atmosphere of a place or situation matters can be heralded as a novel insight. Why then, as Ben Anderson writes, has atmosphere become "good to think with"?[13] As Anderson's own pivotal contribution demonstrates, atmosphere does a number of useful things for scholars in the social sciences and humanities trying to make sense of affective spacetimes. It holds in tension affective spacetimes that are both corporeal and incorporeal, *of* and *emergent from* the midst of bodies, while also having a force irreducible to these bodies.[14] It emphasizes the *relational* qualities of these affective spacetimes as they emerge across and between bodies as dynamic distributions of feeling generated through what Teresa Brennan calls the "transmission of affect."[15] Atmosphere is also useful because it is not restricted to any particular scale: it names a spacetime that can be relatively contained (in a room, for instance) or massively distributed (across crowds, cities, economies, or indeed planets). Regardless of their extent, atmospheres are sustained by the manner in which different bodies, human and nonhuman, move and respond to the conditions in which they find themselves. And they show up through how something of these bodies is expressed, whether as light, heat, color, sound, or gesture. Critically, they have a quality that exceeds any sense in which this experience might be understood as self-contained, or personal. To return to Anderson: atmosphere names the "indeterminate affective "excess" through which intensive space-times can be created."[16] Atmosphere, in short, provides a way of naming diffuse affective fields registering in the sensory capacities of bodies without necessarily being reducible to those capacities. This excess does not, however, mean that atmospheres are beyond intervention. Indeed, they are becoming what Anderson calls the "object-target" of, and condition for, an expanding repertoire of practices and technologies, operating across the domains of the economic, the political, and the sociocultural, designed to generate, amplify, and modulate experiential value.[17] In this context, securing forms of life and the worlds they sustain is a matter not only of maintaining the integrity of territories but also of modulating atmospheres as fields within which certain dispositions to act can emerge or can be prevented from emerging.[18]

As scholars such as Peter Adey, Timothy Choy, Sasha Engelmann, and Tim Ingold remind us, atmosphere is also important because it furnishes a concept for linking the affective with the meteorological: it draws together questions of affect, emotion, feeling, and mood with a concern for the airy, elemental milieu in which entities are enveloped.[19] This milieu is a turbulent and layered envelope around the earth, sustaining different forms of life, and subject to variations of limited calculability operating at a range of scales and over various time horizons. Rendered explicit and potentially governable via various scientific practices, this envelope is increasingly contested as the effects and affects of a range of political-technological interventions are generated, distributed, and resisted, and at different scales.[20] The significance of meteorological variations, processes, and events in this atmosphere is being ever more scrutinized as their origin is denatured, and their disruptive effects become part of the turbulent, emergent urgency of the affective life of contemporary political ecologies.[21]

But meteorological variations are not reducible to the calculative and metrological imperatives of atmospheric science, however dominant this science has become in framing knowledges and politics of the atmosphere.[22] The atmosphere, as an elemental envelope in which life takes place, also matters through the shape of what Tim Ingold has called "weather-worlds."[23] Meteorological variations are felt in domains of experiment and experience that exceed technologies of measurement. They are expressed through changes of color, temperature, and wind speed, and through the felt experience of these changes in ways that resist any neat ontological division between the material and the immaterial, or between surface and sky. Seen thus, the atmosphere is not so much a zone apprehended from a distance but an elemental condition in which bodies are enveloped. Its spatiotemporality is not only meteorological in a climatological sense but also because it shapes senses of feeling or being enveloped in a process akin to an ongoing precipitation of percepts and affects.[24] Ways of describing variations in the meteorological atmosphere—evaporation, condensation, and precipitation—also become ways of tracing modifications in affective experience.[25]

The capacity to sense or be affected by these variations is by no means restricted to the human but extends to nonhuman forms of animal life, and to myriad more-than-human agencies. These variations are sensed when a tree bends in response to wind, or in the slow cracking of rock exposed to the elements. Leaves and rocks may not be moved by atmosphere in the sense of phenomenological experience, but they are affected or "perturbed" by its variations.[26] To make this claim is to refuse what Alfred North Whitehead calls the "bifurcation of nature" into something that, on the one hand, can be sensed as an object of recognition (the atmosphere as a scientific object), and, on the other, that remains as a vague feeling of relation exceeding such recognition (the atmospheric as affective variation).[27] To foreground atmosphere is to refuse to accept the affective and the meteorological as two separate domains, the first pertaining to forms of animate life, and the second existing prior to those forms of life. Insofar as much (if not all) of nonmarine life is enveloped in a gaseous atmosphere, the conditions for affective variation are already meteorological in this sense.

Avoiding any strict division between the meteorological and affective is even more important if we consider how the technical capacity for experimenting with atmospheric envelopes is being extended in all kinds of ways. Certainly, the infrastructures upon which forms of life depend are arguably becoming more atmospheric; that is, they are becoming more ambient, diffuse, and mobile, operating in ways that are responsive to currents and eddies of affective interest while also at the same time generating new inducements to move and be moved that operate below thresholds of conscious attention.[28] Indeed, and notwithstanding the fact that they remain reliant on earthbound systems and sites, the popular adoption of metaphors of the "cloud" suggests that the way in which data, media, and their experiential possibilities are understood is framed increasingly frequently in elemental terms as an envelope of meteorological and atmospheric phenomenon.[29]

The elemental condition of atmospheric envelopment does not pertain to a distinctive ontological or material domain; it is, rather, the force of whatever generates a variation and disturbance that can be sensed in bodies of different kinds while remaining excessive of these bodies. This sensing takes places as the vague, nonrepresentational mode of what in his work Whitehead calls "prehension," a mode of relation that does not need to be understood in terms of the relation between entities.[30] The atmospheric is sensed as something happening that does not need to be an entity; this is something that, as Jane Bennett puts it, "is as much force as entity, as much energy as matter, as much intensity as extension."[31]

ATMOSPHERES BECOMING THINGS

The irony should, of course, be obvious: even if the narrator in Barthelme's story insists that no meaning is to be found in the presence of the balloon, here it has become a lure of sorts for speculating about atmospheres. However, Barthelme's balloon generates atmospheres, and becomes a lure for speculating about the atmospheric, precisely because it is *not* an atmosphere. It is not diffuse, invisible, vague, or ambient. It is what the narrator calls a "concrete particular," hanging there. It is an entity—unified and relatively contained—presenting a surface and retaining an interior withdrawn from view. Because of this, as a device for doing atmospheric things, Barthelme's balloon also foregrounds the perplexing nature of entities as alluring extrusions into worlds whose essence and relations are always beyond us, entities entangled with other entities in ways not always and not necessarily dependent upon human life. The presence of the shadowy creator of the balloon notwithstanding, its form can be grasped through its relations or "intersections" with other things independent of human direction and intervention. The narrator puts it thus: "Each inter-section was crucial, meeting of balloon and building, meeting of balloon and man, meeting of balloon and balloon."[32]

To dwell upon the relations between Barthelme's balloon and other non-human things is to displace—partially at least—the figure of the human in this story. It is also to foreground the radically recalcitrant aspect of those things—that is, the aspect of things that does not just always re-main excessive of meaning but always remains unavailable to any kind of "meeting" or "intersection" of entities, no matter how close those entities might be. Because of this, even if Barthelme's narrator finally lets us in on the secret of the balloon's existence, something of its nature remains withdrawn from its relations with the world or its participation in the gen-eration of a distributed atmosphere; there persists a kind of irreducible and nonrelational otherness only ever finally disclosed in this story in the guise of some omniscient being.

This way of responding to the balloon as a lure for thinking is more in tune with accounts of entities, objects, and things that are part of the elaboration of various speculative realisms and materialisms across the so-cial sciences and humanities. Of course, attention to objects and things is not new, and can be traced through a range of scholarly traditions in disci-plines such as archaeology and anthropology.[33] Much of this work shares the commitment of Barthelme's unnamed narrator to refuse to reduce ac-counts of objects or things to their incorporation within, or representa-tion by, human experience. In much of this work there is also a concern to develop relational accounts of entities and objects. In important recent strands of thinking, however, the promise and possibilities of thinking with and about things and objects are taken even further. Specifically, in writing by philosophers and social theorists in the areas of speculative re-alism and object-oriented ontology (OOO), the injunction to foreground the independent existence of entities is affirmed with particular force.[34] In such work, entities are defined on the basis that something of them exists which can never be incorporated by any other entity—human or nonhuman. The upshot of this definition is not only an acknowledgment of the participation of things in shaping realities but also the development of a speculative and determinedly entity-centered account of reality: enti-ties are the key points of ontological departure for this account, rather than process or relation or movement.[35]

The emergence and elaboration of these claims can be traced through the thinking and writing of various figures. However, the work of Graham

Harman serves as an especially influential starting point. Rereading Heidegger's account of tool use, Harman argues that the most important aspect of an object is not its pragmatic use, nor the fact of its relation to other objects: it is the fact that something of it is withdrawn from other objects. No object is ever able to grasp fully another; no object is fully present to another, no matter how intimate their relations; there exists a constitutive and autonomous remainder to an object that is always "more than its effect on or relations with other objects."[36] For Harman, then, an object is defined not primarily in terms of relations or processes but in terms of "that which has a unified life apart from its relations, accidents, qualities, and moments."[37]

Harman's goal is to rescue the object or entity from the philosophical tendency to regard it as the effect of a more fundamental set of processes or relations. This means moving against the grain of philosophies of process and becoming, and reading these philosophies against themselves.[38] His reading of Whitehead, one of the most important of such philosophers, is illustrative here.[39] Harman takes much from Whitehead's concern with the effort to place front and center the question of how entities affect and are affected (or prehended) by one another. He agrees with Whitehead's claim that actual entities are the ultimate sources of reason. And yet he departs from a reading of Whitehead in which these entities are seen to be ultimately resolvable into processes and events, a reading in which events rather than entities are the real "relata" of relations.[40] For Whitehead, it is worth remembering, "being the situation of a well-marked object is not an inherent necessity for an event."[41] Events take place without necessarily taking the form of entities perceivable or prehendable as such by other entities. For Harman it is precisely these claims that make Whitehead *too relational* and *too processual*, even if for others it is his emphasis on relational processuality that makes Whitehead's work so valuable.[42] For Harman, Whitehead undermines or "overmines" objects by understanding them in terms of something more fundamental—processes and events. In the kind of philosophy of which Harman is critical, objects are always destined, ultimately, to melt or dissolve into something else.

The real challenge posed by Whitehead's philosophy is not, for Harman, how to think process but how to account for the gap between all entities. Where others read Whitehead as developing a philosophical logic of process, Harman uses him to underpin a philosophical logic of entification. By this I mean that differentiation can only be grasped in terms of the generation of another entity or occasion (for Whitehead these are the same thing). In this account it is entities all the way down, across, above, along, and so on, in a process of infinite generative regress. Temporally, according to this account, each entity needs to resolve itself into another entity (or occasion) at any moment to the extent that there is a continuous process of entification (a process which itself is a kind of entity) without any account of what might exist between entities in intensive or extensive terms.[43]

My point here is not to be drawn into the finer points of either Harman's or Whitehead's work.[44] Dwelling briefly on their claims helps us to think about the implication of thinking of atmospheres as entities, however. For Harman, nothing exists between entities that cannot be resolved into further entities. For instance, as Harman put it, "when the tree and I somehow form a new link, we become a new object: every relation forms a new real object," albeit one that is ephemeral and transitory.[45] Equally, the kinds of "meetings" or intersections (in Barthelme's terms) formed between the balloon in his story and other objects are, in themselves, objects or entities. And presumably between these entities and those entities whose "meetings" they facilitate is another entity rather than some gap or void or medium, because any gap or void or medium is in itself an entity.[46] In short, Harman's world is populated by enumerable envelopes in the midst of each of which resides something fully sealed, completely enclosed in itself. Indeed, the best way of imagining this world is to think of it as a world of discrete things in the shape, perhaps, of always already inflated and fully sealed balloons separated from each other by an ontological vacuum that can never be crossed unless it assumes the status of an entity.

The logical conclusion drawn from this is that concepts such as atmosphere and ambience are, in the words of Timothy Morton, another thinker working in this vein, fundamentally "untenable."[47] Morton's problem with such concepts lies with the sense of envelopment they conjure up, and the sense of being-in-the-world with which they are associated. He has the same problem with the concept of world, which he considers, in a manner akin to atmosphere and ambience, as mere "mood lighting."[48] We are so attached to such concepts and we hold on to the ethics, politics, and aesthetics that gather round them, argues Morton, because we lack the conceptual resources to grasp a reality in which we are not within *any* kind of connective or immersive milieu. Instead, like Harman, Morton emphasizes the gaps and voids between things and entities as part of an ethico-aesthetics concerned as much with the nonrelational as the relational. However, like Harman, Morton's aesthetics is also one of the sublime, although where for Harman this resides in the brute metaphysical fact of withdrawal, Morton's sense of the sublime is rather more conventionally scalar: it is about entities (hyperobjects) that are simply too big and vast and massive for us to grasp.[49] Morton is content to hold on to the possibility that the meteorological atmosphere might be understood as a hyperobject: however, he does not want to retain the aesthetic baggage of an affective sense of atmosphere because it remains too wedded to a condition of envelopment that invokes a reassuring plasma in which all things float.

As with Harman, there is much to take from Morton's critique of atmosphere and ambience. He reminds us that these concepts need to be tempered in ways that avoid any tendency to privilege presence or immediacy. This is a reminder that to think of atmosphere and the atmospheric is to invoke an absence, a void, something withdrawn, as much as a common plenum of immersion.[50] However, Morton's critique of the aesthetics of atmosphere also needs to be tempered in turn, not least because such concepts remain vital as conditions, thresholds, and object-targets of a range of actions and interventions without regard for their philosophical tenability.[51] Equally importantly perhaps, without any sense of envelopment as a condition of palpability, it becomes more difficult to account for how the force of the atmospheric (as the expression of variations in affective materiality excessive of entities) is felt and expressed in the senses that these entities have of their surrounds. To invoke the atmospheric is therefore not so much to sustain a misguided sense of some kind of affective spacetime in which bodies are bathed; it is to realize that in different circumstances, bodies or entities are affected by the force of something beyond them that is not an entity.

The logic of entification underpinning Harman's and Morton's account therefore proves particularly unhelpful for thinking the atmospheric. Of course, as Ben Anderson and James Ash argue, it is perfectly possible to think of atmospheres in terms of entities.[52] Following Morton, for instance, we might understand the gaseous atmosphere surrounding the earth, especially through its scientific framing, as a hyperobject. Or, insofar as we speak of the atmosphere of a reasonably contained space (a room, for instance), then we might speak of it as an entity in which other entities are enveloped, or from which they might be repelled, or from which they might seek to escape.[53] As far as atmospheres are concerned, however, it is not always necessary and helpful to think in terms of the formation of a new entity. Think again about the atmosphere of "public warmth" as it emerges in relation to the balloon in Barthelme's story. For Harman, we should understand this atmosphere as an entity. But why should we need to *objectify or entify* atmosphere in this way? This would mean that we would need to understand differences within this atmosphere, differences in intensity and volatility, as entities in and of themselves, to the extent that the entity-like status of this atmosphere would collapse under its differentiation into an infinite number of other entities. Nothing, and everything, would be left between.

The term "entity" or "object" serves particularly badly meteorological variations and differences such as rain or wind. As Tim Ingold has argued, rain is not an entity—it is pure yet palpable process.[54] When we speak of wind, we do not invoke an object, or thing, but a process as multiple as the materials it encounters, resisting any "singular interpretation" through ways of "surfacing" and "mattering."[55] More generally, atmosphere is not a passive medium in which entities are immersed but a field of kinetic and dynamic affects fringed by what, following Bergson, Deleuze, and Massumi, can be called the virtual.[56] There is a generative virtuality pertaining to atmosphere; its actual variations are always open not only to each other but also to a yet-to-be-determined future. The meteorological atmosphere is an elemental field of variation always in the process of actualizing through precipitation, condensation, evaporation, and so on. This turbulent virtuality makes it difficult to predict very far in advance. As Peter Hallward puts it,

> To measure the "actual" weather at any given moment in time and space is like taking a static snapshot of a process that itself remains in a state of continuous interactive change. The weather is only ever actual in the passing instant of a present moment, but what determines every such actuality is the dynamic motion of atmospheric forces as a whole. This motion or force itself cannot be grasped simply by measuring a series of actual states of affairs to which it gives rise: as motion or energy, it exists in intensive rather than extensive forms.[57]

The upshot of this is that while it may make sense to think of the atmosphere as an entity enveloping the earth, the meteorological processes that characterize this entity are not themselves entities. Michel Serres is helpful here. As he writes, a "cloud is a cloud, it is not solely an object. A river is not just an object, neither is an island nor a lake. Likewise the noise of the sea."[58] Serres is not juxtaposing here the dynamic materiality of process against the static materiality of objects but trying to grasp what he calls "the multiple" as a "concept that is common to [a] fluctuating background and [a] stable form."[59] This means that both atmosphere and the variations that precipitate within it are multiplicities of which something is always withdrawn. As Serres puts it, a "cloud is an aggregate, a nebulous set, a multiplicity whose exact definition escapes us, and whose movements are beyond observation."[60] Another way to grasp this is, according to Serres, to think in terms of turbulence—or turbulences—as multiplicities with an aleatory quality from which certain stabilities emerge in the pattern of "the torqueing avenues of eddies along the river banks, a flurry of wind, a string of irregular squalls, the wake of rudders and airships, the throbbing pulse of blood in vessels, the clouds overhead and the clouds of Magellan."[61] Thinking in these terms makes it possible to grasp things as patterned stabilities with a relatively unified existence but which are always characterized by an excessive processuality irreducible to the category of entity.

Perceptive readings by scholars such Ben Anderson, Jane Bennett, William Connolly, and John Wylie offer reminders of the value of Serres's thinking for understanding patterned variations in terms of shapes of change that generate relative stabilities rather than entities.[62] Serres encourages us to attend to stabilities or formations that endure or persist not so much in spite of change and becoming but because of it.[63] This process is akin to the space*temps* of meteorological events or processes, to the space*temps* of precipitations, and to the space*temps* of "populations in disorder constituted by storms, winds, snow and clouds."[64]

This does not preclude thinking with things. It means foregrounding the properties and capacities of things as a speculative philosophical method for thinking about the distributed materialism of atmospheres.[65] This approach allows attention to settle on apparently discrete things in order to affirm contestable claims about the relations and processes from which these things emerge. It is an approach, as Jane Bennett puts it, which "toggles" between objects and the relations and systems of which they are part. In doing so it gives "both objects and relations the periodic focus of theoretical attention, even if it is impossible to articulate fully the 'vague' or 'vagabond' essence of any system or any thing, and even if it is impossible to give equal attention to both at once."[66]

ENVELOPMENT

Deploying the concept of atmospheric things is a way of allowing attention to move between things that under certain circumstances appear as entities and under other circumstances can be grasped only as elemental variations excessive of the category of entity. In doing so, I take the atmospheric to be the force of variations in an elemental condition that can be felt and sensed without resolving the origins of this force into the category of an entity. However, while I take inspiration and support from some aspects of Serres's work, I depart from him when it comes to processes that seem to him to contain or limit any sense of the elemental. Serres has little time for processes of envelopment, or for what he calls the "wrapping" of shapes of change; in contrast, my claim here is that envelopment can actually provide a way of thinking about and disclosing the very processes about which Serres is so emphatic. Envelopment is important precisely because it provides a way of linking a condition of being elemental with the process of fabrication. It provides a way of tracing a process that gives shape to things with the capacity to sense atmospheres but does so without containing their force.

In general terms, envelopment is a process at the heart of which is a certain relation of matter to its ongoing differentiation. We can think of this relation in terms of what Gilles Deleuze calls "the fold," an abstract line along which matter is inflected.[67] The fold is transversal to many techniques and technologies of world making; in envelopment, however, it names a process of partial enclosure through which the relation between distinct phases of matter generates a sufficient degree of difference such that patterned stabilities take shape as variations in that matter. Importantly, the process of envelopment generates entities with the capacity to modify their exposure to variations in the conditions in which they are immersed and around which they are enclosures. Such entities can be grasped, therefore, as capacities to affect and be affected emerging through processes of envelopment that work to modify the exposure of these entities to the conditions in which they find themselves.

Understood in these terms, envelopment is a process through which an atmospheric milieu is modified through a form of enclosure that generates different spheres of life that are inhabitable or uninhabitable to varying degrees. Indeed, in many instances life often depends upon the establishment and maintenance of an enveloping membrane that serves to separate and to some extent insulate an entity from the elemental conditions in which it is immersed. At the same time, envelopment is in many cases a process that allows forms of life to maintain a degree of exposure to these conditions—exposure, for instance, to heat, light, food, air, water, and so on—with which some relation of exchange can be maintained within acceptable limits. The complexity of forms of life (on the surface of the earth at least) can be grasped in terms of the sophistication of the techniques and technologies they employ for maintaining and modifying the relation between envelopment and exposure.[68] This often takes place through the fabrication of secondary envelopes: nests, dens, webs, cocoons, houses, mounds, clothes, and so on. Architecture is the name we give to this process. And architectural envelopment is the process of what Peter Sloterdijk calls "spherification," through which the gaseous and the affective are modified in order to generate environments and conditions in which bodies are immersed to different ends.[69]

This book is not about architecture per se, but it does engage with experiments in which, in the form of the balloon, envelopment generates architectural situations and experiences. Like the more general development of architecture, the emergence of the balloon does not begin in the late eighteenth century. Instead, the balloon is one of the forms taken by the iterative elaboration of technologies and practices of envelopment that generate atmospheric differences. The balloon is, first and foremost, a fabric envelope for the partial capture and containment of a gas of one kind or another. To make a balloon is to fabricate an envelope that, through establishing a relation of difference within an atmospheric milieu, generates a tension between the discrete and the diffuse, between an entity and that which is excessive of the category of entity. This is about modifying the relation between what Michel Serres calls two phases of matter—the gaseous and the textile—in order to produce a new multiplicity.[70] It is about modifying relations within an atmosphere by working with the qualities and properties of fabrics, membranes, skins; it is about working with "veil, canvas, tissue, chiffon, fabric, goatskin and sheepskin . . . all the forms of planes or twists in space, bodily envelopes or writing supports, able to flutter like a curtain, neither liquid nor solid, to be sure, but participating in both conditions."[71] While Serres suggests that this is a topological process in which spaces are shaped and reshaped along surfaces, he also points us to how this is rather a more mundane, pragmatic process. It is a process of experimenting with and testing the permeability of a range of materials, from paper and silk in the early days of balloon flight through goldbeater's skin (part of the intestine of the ox) through more recent materials such as latex, polyethylene, and Mylar.[72] Equally, this process is one involving repetitive, corporeal craft by different bodies. The story of the balloon as a process of envelopment is therefore one that is stitched in part along the pleats and folds of different kinds of fabric, and is told along the seams and skins of envelopes through which emerge discrete, fabricated entities whose shape is held in place by differences in density and pressure as much as by structure or scaffolding.

Inflation is the process through which envelopment becomes shaped. This is not specific to the balloon, of course, nor did it begin with late eighteenth-century experiments with this device. Indeed, in some respects inflation names the very process through which space and time as abstract universals take shape as such. Stories of inflation are stories of the universe—of its shape, size, and rate of expansion or extension.[73] Stories, indeed, of the very shape of reality in process, which, following Henri Bergson, might be understood as "global and undivided growth, progressive inventions, duration," a processual becoming that resembles "a gradually expanding rubber balloon assuming at each moment unexpected forms."[74] At a different scale, however, the relation between inflation and envelopment gives shape to the dynamic shaping of bodies in relation to the atmospheric conditions in which they are immersed. More prosaically, the relation between inflation and envelopment is revealed through the process of breathing. Breathing is an ongoing process through which forms of life sustain a relation of dependent exposure to an atmosphere by enveloping part of that atmosphere for a limited duration. Understood thus, lungs are folds in flesh that allow this process to become one of energetic exchange. Of course, the dynamic relation between inflation and envelopment is not limited to bodies that move through or in the air. In marine animals such as fish, inflation and deflation modify the relations between bodies and water through the swim bladder.

These processes are neither willed nor the product of human intentionality. We did not develop our lungs, nor do we determine the fact of our breathing, even if we can modify it. However, inflation can also become a technical and aesthetic process for experimenting with the relations between devices, bodies, and their atmospheric surrounds. In the case of late eighteenth-century experiments with balloon flight this becomes especially obvious. Such experiments used gases whose generation, nature, and properties were only partially understood.[75] The Montgolfier brothers, Joseph-Michel and Jacques-Étienne, experimented with the combustion of substances such as wool and straw in the hope of discovering the optimal lifting gas whose essential ingredient they believed to be something called "levity." Jacques Charles, their key rival, in contrast, chose to inflate his balloons with hydrogen, lightest and most abundant of all the elements; profoundly affective, solicitous, and with the capacity to combine, often explosively, with almost anything; a gas whose existence had been acknowledged for centuries but had been recognized as a discrete element only in 1766 by Henry Cavendish, and named by Antoine Lavoisier only in 1783. In time, other gases would be used, such as coal gas (a combination of hydrogen, methane, and carbon dioxide generated from the distillation of bituminous coal), first used by the English balloonist Charles Green in 1821, and helium (inert, colorless, and odorless) from the early twentieth century.

The process of experimenting with the technical relations between these processes—envelopment and inflation—allows the balloon to take shape as a device for doing atmospheric things in a gaseous-meteorological milieu. The balloon becomes a shape of change stabilized to some degree by a fabric membrane that is both enveloped in an atmosphere and forms an envelope around an atmosphere. The shape and form of the balloon depend upon the dynamic relation between the two sides of this envelope in a relation of material continuity that is best described not in terms of the relation between two entities but rather in terms of the relation between an entity and an atmospheric condition that is always excessive of this entity.

QUALIFYING ENVELOPMENT

Envelopment is a process that shapes the relation between forms of life and their elemental milieus. While envelopment is not specific to questions of atmosphere and the atmospheric, it provides a way of thinking through the relation between atmospheres and entities in ways that avoid reducing one to the terms of another. It is the process of partial enclosure through which a difference within and between bodies immersed in atmospheres is generated, a difference sufficient to allow for a relation of mediation to occur. This helps us think about Barthelme's balloon, the origins of which remain shadowy and obscure. It is an entity, on one level, to be sure, but one that is always occurring through a process, a process of envelopment in which shapes of change become generative of sensed differences. But thinking about envelopment also helps us temper the critique of this term made by thinkers such as Harman and Morton, and to some extent Serres. These thinkers arguably misread envelopment, or at the very least underestimate its value as a technical process of fabrication as much as a phenomenological sense of presence. In relation to atmospheres, envelopment does not just name the sense of being in some kind of palpable but intangible milieu; nor is it a process specific to human-centered forms of sensing a surround. And it is not a process that overly constrains capacities to sense or think through the elemental. On the contrary, envelopment is a process for experimenting with these capacities by generating a difference within an elemental milieu that takes shape as an entity conditioned by this milieu.

This is not some naïve affirmation of envelopment, however. This focus on envelopment needs some important qualification. Insofar as it is implicated in forms of life, envelopment is not an abstract process but one whose variation is felt in the shape and duration of different bodies. Another, philosopher, Luce Irigaray, helps us to think about this, while also providing an important qualification of any attempt to offer an entity-centered speculative realism inattentive to how certain kinds of processes generate entities that are marked in all kinds of ways. Throughout her writing Irigaray outlines an ethics of sexual difference via a discussion of envelopment in relation to the bodies and spaces of man and woman.[76] In Irigaray's reading, masculine envelopment takes place as a form of enclosure, or circling, or containment. Conversely, woman both acts to provide an envelope, or place, for man, while also lacking her own envelope or place within which to be enveloped. She exists in the aporia or gap between envelope and thing, while also, argues Irigaray, being exposed to the "groundless" ground of being-in-the-air.[77] And yet Irigaray's critique of envelopment does not mean dismissing the term.[78] She sounds both a cautionary and an affirmative note in relation to the promise and problem of envelopment. She is critical of the idea of envelopment as a form of containment, while also affirming a form of envelopment that is always open to that which is excessive of any envelope; Irigaray calls this air, but we can also call it atmosphere, or, in a more expansive sense, the elemental. Irigaray encourages us to rethink how envelopes and envelopment provide possibilities for remaking worlds through different kinds of ethicopolitical association that open up new forms of cohabitation.[79] To think envelopment with Irigaray, then, is to explore how experiments with this process might provide ways of reimagining spacetimes of encounter and experience, a point to which I return in chapter 8. Despite her enthusiasm for architecture, and her call for more architects of envelopment, Irigaray tells us little, however, about the kinds of technical and aesthetic forms this might take.[80] Here Serres and Sloterdijk offer a little more, insofar as they remind us that the envelopes of sense and experience we inhabit are facilitated by all kinds of technological devices. In *Atmospheric Things*, each of these thinkers offers ways of thinking about how transformations in spacetimes might involve experiments with envelopment as a technical and aesthetic process of creative cofabrication.

IN *ATMOSPHERIC THINGS*, foregrounding envelopment is a way of negotiating the tension between an entity-centered account of reality and an atmospheric materialism. Envelopment, in other words, is a process through which atmospheres become things and things become atmospheric. The balloon figures as the shape of this speculative proposition, but it remains an open shape, subject to all kinds of encounters and forces that pull it in different directions. Thinking with the balloon is a way of staying tensed between the technical, ontological, and ethicopolitical senses of envelopment. Technically, we can think of envelopment in terms of how the gas in the balloon becomes something that can be mediated by the air (as medium) outside the balloon. Ontologically, the shape of the balloon discloses a process through which matter folds in on itself, generating an entity with the capacity to act in relation to its atmospheric milieu. Ethico-politically, the shape of the balloon is a reminder that the process of envelopment becomes qualified in the work performed not just by different entities but also in the forms of work and experience undertaken by differently marked bodies, and under different circumstances.

So while it appears suddenly as a discrete entity, and eludes meaning, Barthelme's balloon can become a speculative device for thinking about and doing atmospheric things. It helps us grasp how envelopment generates atmospheric things given form as discrete entities while always remaining open to that which is excessive of the category of entity. We might think of Barthelme's balloon, no less than any other, in terms of what Michel Serres calls an "exchanger," a device that allows a gaseous atmosphere to pass into, and give volume to, an entity with the capacity to move and be moved by this atmosphere.[81] In the process, envelopment emerges as a way of thinking between the force of atmospheres and the allure of entities, turning both against themselves in ways that generate distinctive powers and capacities to act, to sense, and to move.[82]

2

If envelopment generates a difference that can be sensed, then what kinds of atmospheric sensing does this process afford and enable? This question is simultaneously theoretical and empirical. Theoretically, it is about grasping the terms of the relations between an entity and its elemental milieu. Empirically, in this case at least, it is about experimenting with the promise of the balloon to disclose the qualities and movements of this milieu via the sensory capacities of a range of devices and bodies. Different situations can occasion such a mode of experimentalism, from the avant-garde to the ordinary, the monumental to the mundane, the dramatic to the diverting. Not all of these situations involve leaving the ground, however: most, indeed, do not. They can involve simple acts of release, moments of encounter, and minor experiences of immersion. But those situations in which the balloon does afford ways of being and becoming airborne are worth thinking about, and not only as reminders of how the balloon ushered in new modes of seeing the world from above. They can also precipitate speculation about the prospect and affects of becoming untethered,

FIGURE 2.1 Tethered balloon "Ballon Generali," Parc André Citroën, Paris, August 2016. Photo by Guilhem Vellut. Wikimedia Commons.

of being ungrounded, and of being immersed in an atmospheric milieu. They can encourage thinking about what it means to sense this milieu when it never presents itself fully.

SINCE 2002 A LARGE tethered balloon has operated in Parc André Citroën, located in the 15th arrondissement of Paris. So long as the weather is reasonably calm and the wind moderate, on most days the balloon ascends at intervals of about thirty minutes, carrying about twelve passengers a time to a height of about 150 meters.[1] In May 2007 I spent a day in the park, making three "ascents" in the balloon and lingering a little in between. I was visiting Paris to chase up some loose ends related to the Andrée balloon expedition to the North Pole in 1897.[2] The expedition was Swedish, but Paris and Parisian balloons were central to the material, imaginative, and affective geographies of the enterprise. The expedition balloon, *Örnen* (*The Eagle*), was constructed in the factory of Henri Lachambre on the rue de Vaugirard; one of the expedition members, Nils Strindberg (nephew of playwright August Strindberg) had made his first balloon ascents in the city, undertaking six training flights during March and April 1896. The expedition leader, Salomon August Andrée, had also visited Paris on a number of occasions, once during the 1889 World Exhibition, an event that featured two large tethered balloons among its central attractions.[3]

Tethered or "captive" balloons were prominent features at world exhibitions, expositions, and fairs throughout the nineteenth and early twentieth centuries. At the Paris World's Fair of 1878, an especially large balloon was installed, becoming an object of spectacular attraction, even when not in the air. One observer wrote, "From the Gardens of the Tuileries one sees it when the cable is altogether wound up, and the car touching the ground, towering like a great cupola above the highest point of the Palace."[4] Designed by Henri Giffard, this balloon carried about 35,000 passengers during the period of its operation. With a volume of 883,000 cubic feet, it was one of the largest balloons ever constructed, and would ascend to an altitude of five hundred to seven hundred meters, carrying up to fifty people for about twenty minutes.[5] By all accounts, to ascend in such a balloon was to realize aspects of the promise of the balloon as a device for expanding the scope of sensory experience. Vision was obviously central to this promise. Those who paid to ascend in tethered balloons noted frequently how it allowed the world to be seen anew, from a different aspect, although an aspect always already framed.[6] One individual who ascended in 1878 noted that "from the altitude to which the balloon rose this evening Paris resembled the Indian Shawl at the Exhibition, having for its subject a pictorial plan of the town of Cashmere and its environs."[7] Such descriptions echoed accounts by earlier aeronauts of the thrill of the view from above, of being captivated by a world unfolding cartographically. However tame, tethered ascents offered a paying public the opportunity to consume the experience of elevation as a powerful and privileged combination of visual resolution and corporeal transcendence.[8]

The view from above remains central to the allure of ascension, however tethered: in the Paris balloon, ascension is about admiring the view, about seeing the city laid out beneath. As the gondola rises, a view begins presenting itself, the park becoming maplike, geometrical in aspect, with attention shifting to landmarking, feature finding. As it continues to ascend, it is hard to suppress the desire to take it all in, from all directions, all at once: the desire to get the visual measure of the world.

But the process of taking to the air, and of becoming atmospheric, is never only about vision; it is also about the emergence and elaboration of an array of techniques and experiences of sensing what are often fine, fleeting, barely noticeable differences. Even with the tethered balloon, a feel for this kind of sensing can emerge around the elusive moment of lift. It begins at the moment at which the balloon, in being loosened, feels the force of the milieu in which it is enveloped. It begins rather suddenly, torque-like, ever so slightly rotational: anchorage giving way to tensed instability. It happens as a minor corporeal reorientation; an incremental rebalancing of body space; an unthinking adjustment of feet; a torsional ungrounding. Quickly, almost immediately, this calms, replaced by something steadier yet more unfelt: uplift. The force of 5,500 square meters of enveloped helium held in tension by a steel cable, unwinding slowly from below. And so commences the experience of ascent, one that, save for wind, metalcreak, and audible cablestrain, is largely silent.

The simple experience of ascent and descent, even that as safe and tame as the one offered by the tethered balloon, can lead thinking astray, and in more directions than it is possible to anticipate. What cultural geographer John Wylie calls the "ineffably embodied process" of ascension, even in the absence of redemptive, uphill slog, is generative—at least potentially— of differential affects and divergent lines of thought.[9] Over time, these affects and percepts can continue to return and unsettle the effort to think about the relation between entities and the atmospheric conditions in which these entities are immersed. The afteraffects of these experiences can circulate and reverberate in thought far beyond the site-conditioned circumstances of their encounter.[10] And as they circulate, their force can become amplified through exposure to accounts of other balloon experiences. Such minor experiences can complicate the prominence afforded to vision and the visual in accounts of the modes of sensing made possible by the emergence of the balloon as a device for being and becoming atmospheric.

What emerges in the process is not so much the dismissal of vision, but the placing of vision within a wider array of sensing: an array that, following Gaston Bachelard, we can call "aerostatic."[11] Aerostatic sensing is the process by which buoyant bodies immersed in a gaseous, elemental atmosphere prehend and are affected by the conditions and variations of that medium. The emergence of aerostatic sensing involves both or either the body of the aeronaut and the body of the balloon becoming affective envelopes for sensing the elemental condition of atmospheres: envelopes for sensing the affects and meteorological variations of the gaseous atmosphere as a medium in which those bodies are enveloped. This mode of sensing is not specific to human bodies, but it can and does facilitate distinctive experiences of feeling and moving. Understanding aerostatic sensing is therefore about tracing the different ways in which the balloon becomes a device for translating atmospheres into the sensory capacities of different devices and bodies. In this way the question of what it means to sense the elemental force of atmospheres is displaced, away from the pursuit and privileging of simple immersion as part of a critique of visual distance, and toward the problem of how to sense the conditions of a field never fully present to, and therefore always somewhat distanced from, the entities it envelops.

THE VISION THING

It is difficult to move beyond vision as the primary sensory register implicated in the emergence of the balloon as a device for being and becoming airborne. As Caren Kaplan has shown in her work about the emergence of aerostatic mobility and vision, the development of balloon flight was critical to the elaboration of a way of seeing the world from above—which she calls "the balloon prospect"—that combined the cartographic gaze with a new form of mobile, visual appropriation.[12] Accounts of balloon flights in the decades after the first ascents in the 1780s, and well into the nineteenth century, are replete with descriptions of how the world looked from above. After one ascent, the scientist and balloonist James Glaisher wrote,

> The view was indeed wonderful: the plan-like appearance of London and its suburbs; the map-like appearance of the country generally, and the winding Thames, leading the eye to the white cliffs at Margate and on to Dover, were sharply defined. Brighton was seen, and the sea beyond, and all the coast line up to Yarmouth. The north was obscured by clouds. . . . Railway trains were like creeping things, caterpillar-like, and the steam like a narrow line of serpentine mist. All the docks were mapped out, and every object of moderate size was clearly seen with the naked eye. Taking a grand view of the whole visible area underneath, I was struck by its great regularity.[13]

Insofar as the balloon allowed the earth to be seen from above, it gave rise to a mode of seeing that has since become formalized under the name "remote sensing." Put most straightforwardly, remote sensing is a way of "deriving information about the Earth's land and water surfaces using images acquired from an overhead perspective."[14] Privileging as it does both vision and the verticality of ascension, remote sensing is difficult to affirm in any straightforward way as a technique for sensing how bodies become enveloped by atmospheres. If read through the critical lenses of contemporary cultural and social theory, remote sensing can be rendered problematic as both a technique of distancing and as a practice of technical objectivity.[15] By privileging ascension and elevation as techniques for acquiring information and knowledge about the surface of the earth, remote sensing is aligned with all kinds of geotechnical and geopolitical dreams of control through airspace and atmosphere.[16] Critically, the kind of remote sensing made possible by the balloon facilitated the technical alignment of the aerial gaze with formations of geopolitical power and vision, through, for instance, its use as a platform for military reconnaissance in conflicts such as the American Civil War and World War I. Indeed, the advent of the balloon heralded the development of a series of platforms for aerial observation for a range of different purposes, some more malign than others, from the drone to the satellite.[17]

Even if not always based on direct visual observation, remote sensing seems to privilege vision as the primary sensory register through which geospatial and geopolitical knowledge is generated and presented. It seems to reaffirm the putative authority of particular kinds of cartographic abstraction, working against the grain of arguments about the importance of developing situated and embodied forms of spatial knowledge. It seems to rehearse the privileged distancing of subject from object as part of a technoscientific gaze "from nowhere" critiqued so thoroughly by feminist scholars and others.[18] As aeronaut and scientist James Glaisher put it, "The balloon gratified the desire natural to us all to view the earth in a new aspect."[19] For Glaisher and others, then, balloon flight was a literal enactment of philosophical transcendence, and the balloon was a speculative vehicle for experiments in natural philosophy: it allowed the scientist and natural philosopher to cast off the constraints of earthbound vision. Gilles Deleuze suggests that the platonic philosopher is "a being of ascents; he is the one who leaves the cave and rises up. The more he rises the more he is purified."[20] In the kind of remote sensing afforded by the balloon, then, dreams of transcendence through ascension resonated with one of the foundational narratives of the Judaeo-Christian tradition, and with various strands of Enlightenment philosophy, in which the pursuit of knowledge is a movement upward, an attainment of an ideal elevated position free from the grasp of gravity, from heavy, stultifying thought.[21]

On the surface then, remote sensing appears incompatible with theoretical, empirical, and political agendas concerned with grasping atmospheres as elemental conditions in which bodies are sensuously immersed. To affirm or engage with the sensing made possible by balloon flight would therefore seem to imply a critique of remote sensing by privileging immersion, immediacy, and a relation of continuity with the milieu in which bodies are enveloped and in which they move. It would seem to be about being in and being situated, rather than being above, removed, distanced. The narrative of this critique is not so straightforward, however. The emergence of the balloon prospect, and of remote sensing, was not a smooth linear process of technological mastery or of visual appropriation but was entangled in the incremental development of different representational techniques, from sketching to photography. As Caren Kaplan has so convincingly documented, the balloon prospect did not emerge fully formed, coherent, and all at once: it was assembled, acquired, and clarified over a period of time via a range of practices that allowed the experience of being-in-the-air and being above to make sense.[22] These practices included painting, exemplified in Thomas Baldwin's remarkable *Airopaidia* of 1786, featuring elaborately detailed images derived from the experience of traveling in Vincenzo Lunardi's balloon. Photography further enhanced the capacities of the balloon as a platform for elevated image capture. Figures such as Gaspard-Félix Tournachon, otherwise known as Nadar, ascended in the balloon to capture bird's-eye photographic views of cities.[23] These were not so much views from nowhere: rather, no less than on the ground, the experience of moving and seeing from above exemplified the emergence of a form of what geographers David Clarke and Marcus Doel call "vernacular relativity."[24]

It was not only the view of the ground from above that fascinated those who were airborne. Being aloft, and being immersed in the elemental atmosphere, involved attempts to make visual sense of the atmosphere as its shapes and forms varied according to altitude, temperature, moisture content, and wind speed. It involved making sense of scale and distance in a realm with few familiar features or stable reference points, in which the eye could easily be deceived by qualities of light mingling with water in different ways. In this context, to follow John Wylie, the visual experience of being-in-the-air, or being-in-the-atmosphere, like the experience of being-in-landscape, emerged less as a process of gazing upon than as a way of *seeing-with* that is always partial, fragmentary.[25] This way of seeing-with emerged in forms that were as dreamlike as they were about somber technical description. The author, astronomer, spiritualist, and balloonist Camille Flammarion captured something of this in his description of a balloon flight: "Here the most magical panorama which fantastic dreams could evoke presents itself to our contemplation. The central district of France spreads itself out beneath us as an unlimited plane, as rich in colour as varied in tint, and which I can only compare once again to a magnificently painted geographical map. The space around us is of the most perfect transparency. In the midst of these blue heavens I rise from my seat, and leaning my arms upon the edge of the car, I glance downwards into the immense abyss."[26] What emerged, then, with the development of practical balloon flight, were modes of visual sensing in which clarity was always partial and occluded, and in which vision mingled the position of the scientific observer with the projections of the seer.[27]

BEING IN DISAGREEMENT

Another reason to temper a critique of remote sensing follows on from the observation that vision is but one mode of sensing facilitated by being-in-the-air. To really appreciate this point it is necessary to begin not with the sensory capacities of the body of the aeronaut but with the body of the balloon as the shape of the process of envelopment. Envelopment generates a relation of difference within the meteorological atmosphere, taking shape as an entity with the capacity to respond to variations in this atmosphere. We can understand this responsive capacity as a kind of sensing. Clearly, while fabricated by humans, the balloon envelope responds to the conditions of its immersion in ways that are not necessarily dependent upon human experience. It does not necessarily sense the force of atmospheric variation because we feel it do so. It responds to its milieu through what Alfred North Whitehead calls "prehension."[28] For Whitehead, all actual entities or actual occasions prehend their surrounds through a kind of feeling albeit one that takes place "without a feeler." Following Whitehead then means expanding the domain of sensing to include entities, relations, and processes independent of and prior to human life. It also means thinking about how the capacity to sense the force of atmosphere is by no means restricted to the question of what a human body can do.

However, Whitehead's claim needs qualification. First, it is a metaphysical claim rather than a technical one. While prehension may be characteristic of all entities, Whitehead's work is largely silent on the "specificity of the sort of prehensions made possible by the history of technics."[29] Second, Whitehead's account, as Graham Harman reminds us, is about the relation between entities (or occasions).[30] But the relation between a balloon envelope and its surround is not necessarily best understood in terms of entities. Meteorological variations in the atmosphere are not necessarily grasped as entities—even if it is sometimes correct to speak of relatively discrete bodies of air, defined by temperature or pressure, for instance, within the atmosphere. What the balloon envelope is sensing in many instances is a variation, a gradient, or a change. The balloon, as a shape of envelopment, is therefore an entity with the capacity to sense the elemental force of atmospheric variation.

The buoyancy of the free, untethered balloon exemplifies this capacity. The buoyancy of this balloon is reducible neither to the properties and behavior of the gas with which it is inflated nor to the permeability of the envelope itself: it emerges from their interaction with temperature, sunlight, precipitation, and pressure, such that the balloon remains in a kind of ongoing equilibrium with its surrounds. So while it is possible to speak of the buoyancy of an aerostatic *body*, buoyancy is always a thoroughly relational force, the ongoing prehensive sensing of a relation between an entity and the atmospheric medium in which it is immersed. The balloon envelope can be defined in part, therefore, by the capacity to sense—or prehend—the difference between the gas inside and the gas outside. It *feels* this difference.

A free balloon, however, is too good at sensing atmospheric variations. It is too responsive, at least in a horizontal plane. The process— envelopment—that generates a difference within an atmospheric medium also commits the balloon to agree with this medium. It has a static rather than a dynamic relation with this medium: hence the term "aerostatic." Under normal conditions, a free balloon will always be in horizontal "agreement" with the atmospheric conditions in which it finds itself. The term "agreement" is borrowed from the nineteenth-century Swedish balloonist Salomon August Andrée, who put it thus: "A balloon, which floats perfectly freely in a current of air, moves in exactly the same direction and at precisely the same speed as those of the current of air. This agreement between the current of air and the movement of the balloon makes it useless to provide the balloon with such accessories as sails, rudder, etc."[31] Andrée was merely restating a long recognized phenomenon. In the middle of the nineteenth century, the balloonist Monck Mason observed, "To all intents and purposes . . . a balloon freely poised in the atmosphere may be considered as absolutely inclosed or imbedded [sic] in a box of air; so completely so, that (for example) were it possible to distinguish, by tingeing it with some particular colour, that portion of the atmosphere immediately surrounding the balloon, and in that guise commit here to the discretion of the elements, she would, apart from all fluctuations in the level of her course, continue to bear the same tinted medium along with her."[32] This fact meant that it was impossible to make the balloon disagree with its medium, despite various attempts to do so, including the use of paddles, oars, and draglines of different kinds. Ultimately, however, a free balloon is only dirigible insofar as it can be made to use the movement, direction, and speed of the atmosphere in which it is immersed.

STILLNESS IN MOTION

The almost complete agreement between the balloon and the conditions of its medium has an advantage of sorts: it allows the balloon to become a relatively stable platform from which to attend to the experience of being enveloped by what in the early days of balloon flight was called "atmospheric air." Scientist and balloonist James Glaisher wrote that the balloon "leaves the observer entire [sic] free to note the phenomena by which he is surrounded. With the ease of an ascending vapour he rises into the atmosphere, carried by the imprisoned gas, which responds with the alacrity of a sentient being to every external circumstance, and lends obedience to the slightest variation of pressure, temperature, or humidity."[33]

It was not only the medium in which they moved to which aeronauts attended, however. They also reflected upon how the capacity of the balloon to respond to this medium could be felt, and upon how it was translated differentially into feeling via the sensory capacities of the human body. In *Aeronautica, or Sketches Illustrative of the Theory and Practice of Aerostation* (1838), Monck Mason reflected on this experience of being aloft. Understanding this experience, and the "new field of enquiry" of which it was part, revealed the body as an array of materials with different capacities to sense and be affected.[34] He wrote, "It is necessary to be observed, that the human body is composed of a variety of different materials, of different specific gravities, and endowed with different degrees of sensibility to pressure, or other disturbing causes, to which they may happen to be subject. When these are set in motion all together, by one and the same impelling force, a very considerable disarrangement of their relative positions must ensue."[35]

The experience of being aerostatic was often far less "disarranging" than Mason's comments suggest, however. Horizontally, a sense of movement was almost absent. What struck balloonists including Mason was the profound and unexpected stillness they experienced. The movement of the balloon, so obvious to those observing it from the ground, did not translate into a strong feeling of motion within the bodies of those aloft. Mason put it thus: "Without an effort on the part of the individual, or apparently on that of the machine in which he is seated, the whole face of nature seems to be undergoing some violent and inexplicable transformation. Everything in fact but himself, seems to have been suddenly endowed with motion, and in the confusion of the moment, the novelty of his situation and the rapidity of his ascent, he almost feels as if, the usual community of sentiment between his mind and body having being dissolved, the former alone retained the conscious of motion, whereof the latter had by some extra-ordinary interference been suddenly and unaccountably deprived."[36]

It was the absence of movement—or, more precisely, the absence of anticipated "atmospheric resistance"—that stood out.[37] For Mason, balloon ascent did not mean leaving the body behind through an act of transcendence, but instead involved the disarrangement of "the community of sentiment" between mind and body. It took time to recognize and appreciate this feeling, however: only through repeated ascents could the aeronaut learn "to feel the absence of results, which are in fact only remarkable when missed, and only missed when particularly expected."[38]

Later aeronauts also remarked upon this strange feeling of being becalmed amid motion. The pioneering aviator Albert Santos-Dumont described it thus: "We were off, going at the speed of the air current in which we now lived and moved. Indeed for us, there was no more wind; and this is the first great fact of spherical ballooning. Infinitely gentle is this unfelt movement forward and upward."[39] Santos-Dumont and others were not describing stillness as the opposite of movement.[40] Nor was this stillness created by or confined to the individual subject. This sense of stillness in motion was the ongoing outcome of a mobile, distributed, and processual assemblage: it emerged as a sensed and palpable still point tensed between envelopment and the elemental conditions of atmosphere.[41]

FIGURE 2.2 Alberto Santos-Dumont ascending in gondola of his free spherical hydrogen balloon *Brazil* on July 4, 1898, from the ground of the Jardin d'Acclimatation, Bois de Boulogne, Paris, France. Source: Musee de l'air et de l'espace. Wikimedia Commons.

These early experiments with being airborne and of becoming aerostatic were symbolically rich. They were also undertaken and experienced largely by men, as part of a gendered division of labor (see chapter 8). These experimental experiences can therefore be read through the influence of the cultures of heroic masculinity in which they were performed and pursued. In a psychoanalytic reading of Victorian cultures of risk and adventure, Elaine Freedgood argues that balloon flight afforded a regression, however temporarily, to the "oceanic feelings" of the infant: a state in which a subject has yet to differentially precipitate as a discrete, self-contained individual.[42] The balloon, for Freedgood, became an experiential-aesthetic vehicle for immersion in an atmospheric medium simultaneously objectless and vaguely womblike. We do not need to read the experiences of balloon flight in terms of psychoanalytic theory in order to grasp the point being made by Freedgood: those who traveled aloft, almost exclusively men, were confronted with the disquieting sense of being enveloped in an atmospheric medium whose limits were difficult to grasp, sound, or fathom with any degree of certainty.

Being-in-the-air was therefore defined in part by the difficulty of getting the visual measure of a space whose volume and depth exceeded familiar frames of reference. Monck Mason put it thus: "Undisturbed by the interference of ordinary impressions, [the balloonist's] mind more readily admits the influence of those sublime ideas of extension and space which, in virtue of his exalted station, he is supremely and solely calculated to enjoy."[43] Unsurprisingly, balloonists sometimes resorted to an architectural vocabulary in order to get the measure of space. The sky was often compared to a vault, a sphere, or a dome. Clouds were ceilings and pillars. Wilfred de Fonvielle wrote of how, during a flight with Gaston Tissandier, "the masses of vapour hollowed themselves out into a gigantic vault, rendered brilliantly luminous by the reflection of the sun's rays. It was like a vast tunnel of compact cloud through which we were sailing along in silence. The lower portion, as a whole, was like an immense circular basin, such as that in the Tuileries but twenty times as wide and ten times as deep."[44] James Glaisher wrote that the balloon above the clouds "occupies the centre of a vast hollow sphere, of which the lower portion is generally cut off by a horizontal plane." The architecture of the atmosphere was one in which "we can suppose the laws of gravitation are for a time suspended" and above which is raised a "noble roof—a vast dome of the deepest blue."[45]

The suspension of gravity allowing for such structures and shapes was paralleled with a lightening of the self: to some extent ascent was about leaving weight behind, about becoming unearthly. Ascension and elevation became linked in a way that affirmed levity as existential transcendence. Santos-Dumont wrote that the balloonist seemed "to float without weight, without a surrounding world, a soul freed from the weight of matter!"[46] In *Air and Dreams* the philosopher Gaston Bachelard understands the symbolic significance of these experiences in terms of the figure of what he calls "aerostatic man" in dreams, now largely forgotten with the dominance of aerodynamic flight. For Bachelard, the invention of the aerostatic balloon seemed to substantiate a dream of floating flight that precedes the invention of airplanes. Citing the nineteenth-century writer Charles Nodier, Bachelard writes that "aerostatic-man, the resurrectional man" has a "vast, solid, enlarged torso, 'the shell of an aerial ship.'" He flies by creating "at will, a vacuum in his extensive lungs and by striking the earth with his foot, as the instinct arising in his developing organism teaches man to do in his dreams."[47] Here the fantastical corporal buoyancy of aerostatic man becomes analogous to that of the balloon as it senses its immersion in an atmospheric milieu.

The elemental buoyancy of aerostatic man, of course, has important antecedents. Archimedes, in *On Floating Bodies*, conjures up a sense of the feeling of being immersed in a fluid. In some respects Archimedes offers nothing more than a description of the mechanics of buoyancy: "If a solid lighter than a fluid be forcibly immersed in it, the solid will be driven upwards by a force equal to the difference between its weight and the weight of the fluid displaced."[48] Michel Serres reminds us, however, that Archimedes may also have had a different sense of buoyancy, felt through the movements and displacements of his body. For Serres, the force of buoyancy, as the displacement of a volume, becomes, through Archimedes's body, felt as a kind of joyous experience. He describes it thus: "Here's the naked engineer, in his bath. His floating body undulates, alone, in the volume like a toy boat in a small tub, where his limbs give themselves over to the slight rolling. A naked body, transparent fluid, a theorem of equilibrium by means of the waters: Archimedes feels the force that, by lifting him, causes him to float and swim."[49] In the movement of the body of the engineer, through its corporal joy, the physical force of buoyancy is translated into the affective experience of being buoyant through the elements of air and water. "Archimedes," Serres writes, "felt the movement and rose with the emotion of these two elements, as though he heard the murmur of the waves and the vibration of the wind. And I hear *eureka* as this triple echo of body and air and water."[50]

Dreams of such elemental buoyancy through aerostatic being are often joyous, characterized by elation as the affective vector of elevation. The dream of lightness and of becoming aerostatic, however alluring, is never entirely benign. It can be strange, disturbing, imperiled by an upward pull that can become too strong. Odilon Redon's lithograph *L'oeil, comme un ballon bizarre se dirige vers l'infini* (The eye, like a strange balloon, moves toward infinity, 1882) gives us a glimpse of this. The image appears—at least at first glance—as a surreal rendering of the prospect of transcendence through aerostatic space.[51] All the necessary elements are there: a balloon-like eye, suspended in the air, gazing toward the heavens. There is something disquieting about the heavenward prospect of this strange, aerostatic body, however: something about the immensity of the field through which vision extends. Redon's balloon eye seems untethered yet simultaneously in a steady state of suspension between heaven and earth. Its apparent suspension is no guarantee of a condition of benign equilibrium, however. It presents us not just with the prospect of the weightless, limitless gaze of which aerostatic things are generative. It also preempts the disquieting prospect of becoming too light, of becoming dangerously untethered under the power of the balloon as a "terrifying, inert and baneful thing."[52]

The prospect of this disquietude can be sensed, and is in some respects heightened, in a tethered balloon, where the tension between cable and lifting gas is felt affectively as the *anticipation* of movement—of sudden, rapid, uncontrolled ascent. This feeling lurks as a barely sensed unease immanent to claims for ascension as transcendence. As an affective complex, this is not quite fear. Nor is it vertigo, even if it is vertiginous. It is felt as sensed anticipation in the present of the future event of untethered uplift: a skyward "uprush of the unconscious."[53]

REMOTE SENSING

Variously baleful or benign, aerostatic sensing is never about a feeling of pure immersion, whether of balloon or body. Being and becoming aerostatic, as a mode of sensing the atmospheric, does not provide a more authentically embodied counterpoint to the perceived distancing abstractions of the kind of remote sensing outlined earlier. The dreams of immersion, or what, following Mitch Rose we might call "dreams of presence," that surround the figure of aerostatic being are just that—dreams.[54]

Aerostatic sensing is a modified version of remote sensing: it never escapes the kinds of distancing implied by the impossibility of presence. To think with remote sensing therefore requires modifying its meaning a little. Less a technology of distanced, elevated image capture, remote sensing might be better grasped as the very condition of sensing elemental variation in an atmospheric milieu: that is, the condition of sensing the force of something without direct contact or touch with an entity as such. To understand remote sensing in this way is to return to and modify a more basic yet no less pragmatic definition of the term as "the art or science of telling something about an object without touching it."[55] Remote sensing via satellite or aerial observation might therefore be understood as really only a more technologically sophisticated iteration of a much more basic process of sensing without coming in contact with an entity.

A further modification is required, however, in order to think of remote sensing in relation to the atmospheric: remote sensing involves sensing something that is not resolvable into an object. It is the art of sensing variations in an atmospheric field without coming into direct contact with that field precisely because there is no thing or entity or object with which to come into direct contact. Remote sensing is the becoming sense-able or palpable of that which never has, and never will, precipitate or crystallize as an entity. In the case of remote sensing of the more high-technological kind, what is sensed without being touched directly is a disturbance or modification of the electromagnetic spectrum (for instance the heat radiated by the sun or the earth), which is then registered in a sensing body of one kind or another. In the kind of remote sensing proposed here, what is sensed is also something spectral: the atmospheric is spectral, because it is never reducible to the status of an entity or object whose existence is disclosed fully. Following Derrida, we might say that the atmospheric is spectral because it is the force of that which can be sensed while never resolving into something immediately, fully present.[56] Concepts of atmosphere and the atmospheric are always haunted by something beyond the kind of immersive, immediate presence they promise. To sense the force of the atmospheric is not only to feel immersed in something, it is also to feel the absence of that which is always held back, and in ways that can be variously unsettling or joyous. Insofar as the force of the atmospheric is sensed via a modified form of remote sensing, it never crystallizes fully through what Whitehead calls "perception in the mode of presentational immediacy." Remote sensing remains as a form of "perception in the mode of causal efficacy," as the sensing of variations that are only vaguely felt.[57]

What this means, in turn, is that it is not quite correct to say that an atmosphere can be sensed in a body. This is because so much of an atmosphere always remains beyond the sensory capacities of bodies, always remains withdrawn and remote. It might be more appropriate then to speak of the possibility of sensing the atmospheric as some kind of variation in a medium of immersion. Given this, remote sensing does not name a technique of withdrawal or distancing from a world in which bodies are immersed: it names, instead, the process of sensing variations in milieus whose condition it is to always be partially withdrawn from those bodies possessing the capacity to sense these variations.

A further distinction is worth foregrounding here: between active and passive kinds of remote sensing. In remote sensing of the more obvious, high-technological kind, *active* techniques are those that generate their own source of electromagnetic radiation: directed at an object, this radiation is reflected, with the resultant signal being recorded by a sensing device of some kind. *Passive* remote sensing, in contrast, uses electromagnetic energy transmitted by the sun and reflected from objects. The kind of remote sensing that might be performed in relation to the atmospheric involves a combination of active and passive sensing, of affecting and being affected. This form of remote sensing, as a technique for sensing spectrality, proceeds as an ongoing oscillation between activity and passivity, or, more precisely, as the continual prehensive responsiveness of bodies to elemental variations in the fields through which they move and from which they emerge.

AS A DEVICE FOR sensing, the balloon is not only significant because experiments with it have facilitated distinctively modern practices of airborne vision and visual appropriation. It also facilitates experiments in sensing the atmosphere as a medium in which bodies are enveloped and the atmospheric as the felt variation in that medium. The balloon facilitates the development of a form of aerostatic sensing: this names the particular version of remote sensing that emerges through the development of the balloon as a practical yet always experimental device for becoming lighter-than-air. Rather than offering an experience of pure immersion, this kind of sensing is always conditioned by something withdrawn from apprehension. As a modified version of remote sensing, being and becoming aerostatic therefore never resolves itself into the condition of immediate self-evident presence—of being-in-the-air—that ideas about immersion would seem to suggest. Nor, indeed, is it grasped as a mode of sensing space defined against abstraction and elevation.

To think about the kinds of sensing facilitated by the process of becoming aerostatic is a reminder of the difficulty of privileging of presence within any medium, atmospheric or not.[58] It is a reminder that no entity is ever fully in an atmospheric medium, not because this medium does not exist, but because this medium never discloses itself fully to that entity; instead, it is more accurate to think in terms of how entities can sense variations in atmospheres as the force of the atmospheric. In this respect, any attempt to sense the force of the atmospheric is always haunted by the spectral because the origin of that force will never be disclosed fully.[59] To experiment with and experience immersion in and envelopment by atmospheres through becoming aerostatic is to be tensed between the capacities of things to sense atmospheric variations and the fact that these variations will always remain excessive of these capacities (to some degree). Because of this, the force of the atmospheric is not sensed as something issuing from a discrete source or thing: to borrow from Deleuze, it does not need to be grasped as an "object of recognition" but can be apprehended as a certain affective tonality, a palpable variation.[60]

In this case, remote sensing does not designate a more inauthentic relation between entities and their conditions. This kind of sensing is not remote because it is mediated by a technological prosthesis: rather, sensing, or being affected, is already always prosthetic, a kind of originary technicity of bodies articulated in distinctive ways in human bodies.[61] Sensing is always already remote because it always already reckons with a condition of withdrawal. This means there is never any mode of privileged or authentic sensing from which remote sensing might be seen as a fall from grace. Sensing is not the preserve of the capacities of human bodies. It is spread out and distributed across an array of devices, an array that is becoming more and more complex.[62] To sense the force of the atmospheric through this array is a process that displaces any assumption about authenticity rooted in a human body, while nevertheless remaining attentive to the capacities of this body.

Whatever the device or array of devices, however, the process of sensing atmospheres is never only a technical one. It is also aesthetic. It is aesthetic because it raises the question of the conditions and limits of what can be prehended by an entity. Clearly, processes of being and becoming aerostatic generated distinctive kinds of feeling in the bodies of those aloft, while also prompting reflection on the limits of the capacity to grasp extension, volume, and depth. But far from such reflections on a kind of transcendent sublime, the aesthetics of sensing atmospheres also reside in encounters and experiences that are far more mundane, modest, and minor. They reside in the capacity to sense, or feel, variations in conditions that never become present as such. The feeling of these variations is central to the allure of elemental envelopment, even if it is a feeling that moves along many different affective vectors, variously joyously light and unbearably heavy.

3

The party was only a few blocks away. It was early October 2013 and yet the warmth of the summer never seemed to want to let us go. As we walked, you spotted the yellow balloon tied to the bike, and I took a picture. Your friend, the birthday boy, whom I didn't really know, was three. And in honor of the occasion he had been given a large, helium-filled, purple, foil balloon in the shape of the number he was celebrating. It floated in our midst, a kind of centerpiece for the event. It had been weighted, but, as we were to discover, not securely enough. A little later, as we gathered in the garden to eat cake, someone cried out. Something was happening. We all looked up as it floated out of the garden. Everything was momentarily heightened. The gasp was audible, expressing dismay, feigned or not, at its departure. Perhaps it would not get very far, perhaps it might be recovered. Indeed, it seemed to hesitate briefly over the trees, as if it might fall back to Earth. But it continued on, upward.

You were surprised at how long you were able to track its ascent. You lost sight of it momentarily, but someone else found its shape again and pointed. There! Your eye followed the line to a dot ascending along a near perfect diagonal into the distance. It hadn't simply drifted up. For a moment it appeared to move with direction, as if possessing some sense of its own trajectory. You followed it for as long as you could, for as long as it was polite in company, knowing that as soon as you took your eyes away the thing itself would disappear. And as you did, you wondered: Where do such things end up, eventually? Who or what finds them? What becomes of them? And then you moved on, the party drawing you back in.

IF ATMOSPHERES ARE ALWAYS felt as the force of something withdrawn or never fully present as an entity, then how to account for their aesthetic pull or gravitational draw? Philosopher Graham Harman provides one answer to this question when he names the aesthetic dimension of the withdrawal of things their "allure." The concept of allure underpins Harman's object- or entity-oriented account of the world by highlighting the radically distributed nature of a kind of nonrelational metaphysical aesthetics.[1] Allure names the fact that an object (or entity) cannot present itself fully to another object: something always remains inaccessible to whomever or whatever perceives or apprehends it. As Steven Shaviro, commenting upon Harman's work, writes, an "object becomes alluring precisely to the extent that it forces me to acknowledge this hidden depth, instead of ignoring it. Indeed, allure may well be strongest when I experience it *vicariously*: in relation to an object, person, or thing that I do not actually know, or otherwise care about."[2] For Harman, however, allure not only describes relations between humans and things, it describes how all objects or entities, human or nonhuman, relate to one another, and indeed to themselves and their own variation, in a way that never resolves itself into copresence.

Harman's account of allure provides an important point of departure for thinking about the aesthetic dimension of sensing atmospheres, especially in the light of a revised version of remote sensing. It needs modification, however. If we follow Harman, everything is alluring simply because it is a distinct and unified entity of which something is always withdrawn. But if we take this metaphysical claim seriously—if we take allure to be an ontological quality of everything—then we also need to ask if there is anything special or distinctive about the allure of different things. Is a balloon any more alluring than, say, a spoon, an anvil, or a toilet brush? Harman does not offer us a way of thinking about this: his claim about allure is a metaphysical or ontological one, and is not specific to any entity in particular.

Things *are* alluring in different ways, however. In certain circumstances, and often by design, an anvil may be far more alluring than a toilet brush, and vice versa. To think about the allure of things can never only be about making a metaphysical claim about the fact of withdrawal. It also always involves attending to the specificity of allure in relation to particular things, and, in certain cases, attending to the processes and practices through which allure is fabricated. In this way, as Steven Shaviro suggests, Harman's aesthetics of the sublime can be supplemented with an aesthetics of the beautiful based upon what can be seen, about what is visible, or about what presents itself as a surface of captivation: or, put another way, any account of allure needs to grasp how surfaces are as alluring as whatever remains hidden in the shadowy depths of essence.[3] But Harman's discussion of allure also needs to be modified in a second way. Allure is not the preserve of entities. That which is excessive of an entity—in this case the atmospheric—can also be alluring, both in the sense that it is withdrawn, and in the sense that the mode in which it becomes present is aesthetically alluring.

This doubly modified sense of allure offers a way of thinking through the relation between the material process of envelopment and the aesthetics of atmospheric things. Envelopment provides an important technical process of material fabrication with which to think through, generate, and experiment with the allure of atmospheric things. With respect to the balloon this takes place most obviously through the technical achievement of ascension and transcendence: the balloon envelope is an entity whose allure has much to do with its capacity to withdraw from tangible or palpable presence by rising into the atmosphere. But the allure of atmospheric envelopment also involves the fashioning and fabrication of surfaces that present different qualities of color, shine, reflectivity, and feel. The allure of atmospheric envelopment is about the fabricated capacity of different entities to participate in the generation of what, following Nigel Thrift, we can understand as worlds whose surfaces are shaped by the aesthetic force of captivation, worlds in which almost "every surface communicates."[4] The allure of the atmospheric must also be grasped, therefore, in terms of how it is actively fabricated and engineered through the properties, qualities, and surfaces of envelopes as they become atmospheric things and things as they become atmospheric.

The development of practical balloon flight in the late eighteenth century may have been about the emergence of new ways of viewing the earth from above, and of sensing the atmosphere as a wondrously perplexing condition of immersion. But for most people—in fact for the vast majority—the advent of this technology of travel was actually about the view from below: it was about the prospect of an object ascending into the heavens; and it was about the possibility of following the progress of that ascent until the object was no longer visible. This was an early instantiation of what Fraser MacDonald calls the "perpendicular sublime": the embodied vision of the spectator was drawn upward, in thrall to the ascent of a technical device that for the very first time had the capacity to disappear into the sky.[5]

The development of practical balloon flight was therefore arguably as much about the generation of atmospheres of captivation on the ground as it was about the development of new ways of sensing atmospheres aloft. Put simply, given the novelty of a balloon, the prospect of its ascent drew a crowd, and the gathering of this crowd generated a distributed affective atmosphere of occasion. During August 1783, in Paris, Benjamin Franklin witnessed and wrote about early balloon ascents. He described his experience in a letter to a friend in England, the scientist Joseph Banks. As Franklin described it, after the launch the balloon "Diminished in Apparent Magnitude as it rose, till it enter'd the Clouds, when it seem'd to me scarce bigger than an Orange, and soon after became invisible, the Clouds concealing it. The Multitude separated, all well satisfied and delighted with the Success of the Experiment, and amusing one another with discourses of the various uses it may possibly be apply'd to, among which many were very extravagant. But possibly it may pave the Way to some Discoveries in Natural Philosophy of which at present we have no Conception."[6]

As Franklin's account suggests, the allure of these early ascents can be explained via an appeal to a sense of curiosity and wonder about the public demonstrations of the technical achievement and power of science.[7] There was also, it has to be said, something quasimagical or mystical about ascension. From the very beginning, however, the allure of balloon ascent was also something to do with the sense of occasion it generated. Equally, from early on, this occasion could be and was commodified as an atmospheric event. Even if the balloon aloft provided, as it were, free-to-air entertainment, ascents were opportunities to commodify atmospheric experiences defined by the circulation of affects across and between bodies. Part of this was necessity: balloon launches were expensive enterprises, and early aeronauts, not usually wealthy, used public subscriptions to fund their speculative excursions.[8] And what they could sell was an experience: the experience of witnessing and viewing the launch, or the privilege of seeing the balloon in an enclosure from which it was sheltered both from the elements and the public gaze. The balloonist also traded on the promise of a sense of occasion generated by a hinterland of attractions that often gathered around the prospect of the ascent, through entertainments staged before, during, and after the launch, including, for instance, fireworks.

From the outset, then, balloon ascent exemplified a much wider process for generating the "singular quality of allure through the establishment of human-nonhuman fields of captivation" as part of the emergence of new and distinctively modern experiences and atmospheres of consumption.[9] The alluring atmospherics of experience that gathered around balloon launches was shaped, in turn, by the commodification of accounts of these launches. And publicity, the storied mediation of the promised event, was crucial to amplifying the anticipation that gathered around and funded these enterprises. For the balloonist, newspapers, advertisements, and pamphlets were central to generating collective interest in the prospect of a launch. To those interested in selling a tale, a balloon story, real or invented, promised an audience. Relationships with print media were cultivated, advertisements taken out in newspapers, pamphlets published.[10] Stories of balloon launches and aerial excursions could provide sensational news copy. As Michael Lynn writes, "Commercial advertising of balloons played a significant role in the eighteenth-century press; the press flooded the market with almost daily news reports on this invention, those involved in its development, and details of all its launches."[11]

The balloon very quickly became a vehicle for speculative showmanship. For instance, in 1785 the Italian aeronaut Vincenzo Lunardi undertook five balloon excursions in Scotland, including two in Glasgow. Such excursions were a novelty in late eighteenth-century Scotland: the first demonstrations of practical balloon flight had taken place in France only two years earlier. Large crowds paid to view the inflated balloon on the ground and to witness its ascent. The *Glasgow Advertiser*, one of the sponsors of the event, reported that "upwards of 100,000 spectators assembled" to witness the launch, "among whom were the greatest number of Ladies ever seen in Glasgow."[12] Lunardi, like many aeronauts, knew how to mix chemistry and charisma. He knew that intensity of anticipation could add value to the promise of an event. And he had learned from experience that a crowd gathered in anticipation is an affectively volatile collective. A year earlier, Lunardi had made one of the first aerial voyages in England, from St. George's Fields, London—an event attended by up to 150,000 spectators.[13] His London ascent followed earlier, unsuccessful launches by other aeronauts, the failure of which generated enough anger among the assembled crowds to threaten the safety of both balloon and pilot. Lunardi's attention during the hours before his London launch was therefore divided equally between the inflation of his balloon and the mood of the crowd. Realizing the former would not be completed by the appointed launch time, he made a quick decision, designed to calm the assembled spectators: he would begin his ascent *before* his balloon had been inflated fully with hydrogen, and would ascend without his full complement of advertised passengers. So, as he wrote, "at five minutes after two, the last gun was fired, the cords divided, and the Balloon rose, the company returning my signals of adieu with the most unfeigned acclamations and applause. The effect was that of a miracle on the multitudes which surrounded the place; and they passed from incredulity and menace, into the most extravagant expressions of approbation and joy."[14]

Lunardi's balloon was clearly not only atmospheric insofar as it allowed the balloonist to experience distinctive feelings of being-in-the-air. It was always already atmospheric on the ground: it was atmospheric in the sense that it generated a collective field of affectively felt variation. On the ground prior to launch, it gathered an affective atmosphere, promising a palpable sense of occasion organized around the prospect of an event to come. Then, in the act of its release, in moving skyward, the passage of the balloon became an affective vector: its buoyancy modulated and drew together the affective intensity and tone of an atmospheric event, reorienting, in the process, the bodies of those generatively immersed in this event. To look up was to feel—to remotely sense—the alluring force of uplift fabricated through captivating envelopment.

SURFACES OF CAPTIVATION

The significance of the balloon as a technology of captivation extended well beyond the public spectacle of carefully staged launches. Its shape began to circulate in the form of a range of objects, baubles, and materials. In the months following the first launches, the balloon became highly fashionable, especially in Paris, its design replicated in a range of popular materials and practices: miniature balloons were all the rage, dresses and shirts were fabricated in the style of the aerostat, and artifacts from clocks to lamps were produced in the shape of the balloon.[15]

The period of such "balloonmania" was comparatively brief; the appeal of the balloon as a relatively cheap object of amusement and captivation endured, however. Indeed, as the nineteenth century progressed, the balloon became more and more associated in the public eye with amusement and entertainment than with scientific advance and technical achievement. This is evident, for instance, in how the balloon figured in genres of popular entertainment, and in widely read fictional narratives throughout the nineteenth century.[16] In these stories it was easy to send up the worthy ambitions of serious balloon excursions. In *Tom Sawyer Abroad*, for instance, Mark Twain narrates a farcical trip in an apparently dirigible balloon owned by a balloonist who laments the fact that neither he nor his craft is taken seriously any longer. As they drift over the United States, Tom Sawyer and Huckleberry Finn wonder about their whereabouts, confused because the colors of the country beneath them do not match those they have seen on classroom maps.[17] Published initially in a special supplement to an edition of the *New York Sun* from April 1844, Edgar Allen Poe's "Balloon Hoax" also sent up the worthiness of the balloon. Sold on street corners by a network of newsboys, the *Sun* was a penny daily, the most successful example of a new kind of newspaper whose emergence, growth, and circulation had been made possible by advances in print technology. Poe understood the power of these newspapers to circulate stories and their affects. He also understood the importance of the formal properties of the successful hoax.[18] His "Balloon Hoax" sent up the machinic production of news stories whose affects are consumed by mesmerized subjects. But such was the interest it generated that Poe could not lay his hands on a copy of the newspaper in which it appeared.[19]

In the industrializing countries of Europe and North America, the transformation of the balloon from a wondrous vessel for aerial navigation into a cheap, everyday technology of captivation was facilitated by the development of new materials from which it could be fabricated. In 1824, Michael Faraday first used rubber to make a balloon, albeit for laboratory purposes.[20] In the wake of this development, it soon became possible to purchase kits with which to inflate toy versions of Faraday's balloon,[21] and within a few years such balloons were being sold commonly on street corners by a small army of balloon vendors. Writing of this time, one British balloon distributor noted, "Most of the goods being comparative novelties, I had very little trouble disposing of them . . . we had customers for these balloons, all the vagabonds, street hawkers and costermongers for miles around waiting a long time before the place was open. They actually fought for them, there was such a quick sale. They all had to be inflated by a bellows, coloured powder being blown into them and tied at the end before being ready for sale."[22]

These balloon sellers were marginal figures, on the fringe of everyday urban life, whose wares were minor objects of color, distraction, and amusement. The figure of the balloon seller has frequently been depicted thus. In the poem "In Just," E. E. Cummings writes of a "mud-luscious world" in which "the little lame" and "queer old balloonman" whistles "far and wee."[23] In Fritz Lang's M, a blind balloon seller hears someone whistling a tune which generates a memory of a moment when a young girl was kidnapped and murdered. Lang's balloon seller, in turn, is echoed in the appearance of a similar figure in Carol Reed's The Third Man. Most recently, in Disney-Pixar's Up, the balloon seller (Carl) is a grumpy, crotchety character whose irritability is at odds with the consumer objects with which he has lived his life. The reality, of course, is that in many other contemporary contexts, selling balloons on street corners remains a way of making an income for those whose living conditions are in some circumstances closer to those nineteenth-century hawkers than to the character of Carl in Up.[24]

In the ever more complex and choreographed consumer worlds of the nineteenth century, the balloon was deployed as a device for enhancing the atmospheres of ambient captivation that gathered around commodities and technologies.[25] More than this, they contributed to the commodification of the surfaces that shaped the allure of these worlds. The lightness of the balloon—both in the sense of weight and illumination—made it a particularly promising device and vehicle for advertising.

Patent applications provide one way of revealing the history of attempts to render the balloon envelope as a surface of captivation as part of an emerging set of technologies and devices for generating alluring atmospheres. For instance, in 1873 William Frank Browne of New York filed an application for an "Improvement in Balloon Advertising," consisting of a balloon filled with hydrogen gas, beneath which was to be suspended a large netted surface to which letters could be attached.[26] Robert Wilson, of Memphis, Tennessee, filed an application in 1893 for an "Advertising Balloon," which would use a combination of illuminated transparencies

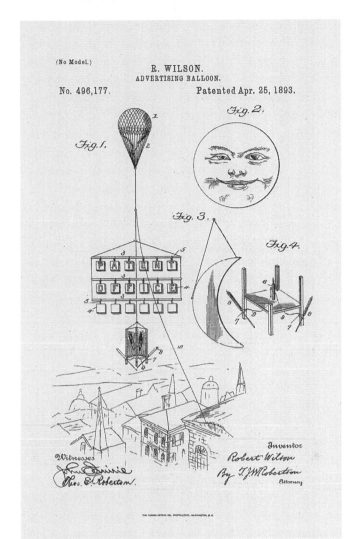

FIGURE 3.1 Diagram of patent application for advertising balloon (1893) by Robert Wilson. Patent no. US496177A.

and fireworks to attract attention, hopefully with no hydrogen involved (figure 3.1).[27] A patent filed by Arthur Selzer in 1917 described an "illuminated toy balloon and lighting effect." According to Selzer, the invention had "for its object to provide such a device which will be particularly suitable for parties, celebrations, advertising displays, etc., and in which a balloon having a handle may be so illuminated as to present a very attractive display."[28] In 1924, J. Eustace Maguire filed a patent for a "Balloon Advertising Method and Apparatus," the purpose of which was "to provide a method by which advertising circulars, novelties, or confetti can be sent aloft in the air, preferably over the heads of a multitude such as might congregate at any public or social gathering, and then at some opportune or predetermined time scatter the circulars or confetti over the people."[29]

From early on in its life as consumer object then, the balloon provided a device through which to enhance the atmospheric allure of commodities and their worlds. In the twentieth century the balloon and the consumer events in which it was deployed became ever more elaborate and choreographed. Significant in this respect is the use of balloons or balloon-like figures by corporations involved in the manufacture of one of the materials—rubber—from which it was fabricated. Michelin's Bibendum, known in English as the Michelin Man, is an important antecedent here.[30] While it made balloon tires, Michelin, unlike its German rival Continental, did not supply rubber for the fabrication of aerial balloons. Nevertheless, both companies had balloons emblazoned with their names displayed prominently at the 1909 Air Show held in the Grand Palais in Paris. Another rubber company, Goodyear, launched its first blimps in 1925 as corporate icons. At the 1933 Century of Progress Exposition in Chicago, two of the blimps carried illuminated boards that flashed messages at visitors below. As prominent "signs in motion," Goodyear blimps have since become associated with large sporting events, most notably the Super Bowl.[31] A Goodyear blimp was also part of the staging and choreographing of the atmospherics around the 2012 Olympic Games in London, its presence in the sky drawing together levity and festivity with the promise of security. More generally, the advertising blimp discloses the possibility of rendering aerial envelopes as moving surfaces of captivation, a possibility realized in cinematic form in Ridley Scott's *Blade Runner* (1982). In that film, blimps feature as enormous looming aerial billboards advertising the promise of a life "Off World" as they move slowly through the cluttered airspace over a dystopian Los Angeles. The surface of the balloon envelope is obviously also deployed as part of the generation of atmospheres around commodities and corporations that are not associated with the manufacture of materials such as rubber. Perhaps the most iconic example of this is the annual New York Thanksgiving Parade, first held in 1924 as a publicity initiative by Macy's Department Store. The central attractions in the parade are enormous, elaborately designed helium-filled balloons, the first of which was Felix the Cat.[32]

More generally, of course, the balloon is deployed frequently in a range of modest ways, from product launches to store openings, to add allure to events, occasions, and commodities. The atmospheres that gather around these entities and events are central to the generation of consumer worlds. In the process, surfaces of all kinds become part of the fabrication of allure. Indeed, as Nigel Thrift has argued, one of the most distinctive aspects of the engineering of allure is the activation of the qualities of surfaces.[33] In this respect, the appeal of the balloon has been amplified by the development of more recent, and more synthetic, "glamorous materials" as part of the revolution in plastics that took place during the period following World War II.[34] Companies like DuPont were at the center of this revolution, and they dreamed up visions of new ways of living made possible by these materials. Lightness, strength, and durability were at the heart of these dreams. In 1952, DuPont produced a twenty-minute promotional film about a new plastic it had patented called Mylar. Much of the film emphasized the strength, durability, and flexibility of Mylar, and the narrator emphasized the aesthetic qualities of this polyester material. He began, "Because of its remarkable versatility, it can be used to obtain effects that are both beautiful and practical." Two female assistants displayed and draped roles of the film, while the narrator continued, "It prints beautifully. It can be metalized. It can be laminated, or bonded to wood, metal, cloth, paper, leather, plastics, and a variety of other materials. It can be coated. It can be fabricated, stamped cut, or spiral wound, or it can be formed under heat and pressure under a variety of designs. Mylar's versatility, and its properties, suggests possibilities only limited by our imaginations."[35] Crucially, as a plastic film, Mylar could be coated with a metalized foil to produce a material of tremendous reflectivity. This process of metalization meant the balloon envelope could become even more alluring: allowing it to present a shiny surface of captivation in which the world around it could be reflected.[36]

ATMOSPHERES OF IMMERSION AND RESPONSIVE ENVELOPMENT

Peter Sloterdijk has argued that the generation of atmospheres through the engineering and fabrication of architectural envelopment is central to the production of the commodity worlds that shape modern spheres of inhabitable life.[37] The qualities of the materials used in this process are obviously crucial. Buildings such as the Grand Palais in Paris, or the earlier Crystal Palace in London, employed advances in steel and glass in order to produce structures defined by an immersive atmospherics in which lightness, both in illumination and in materials, was transformed into a feeling of soft spaciousness: a feeling of being both inside and outside, of being enveloped without being contained. As one visitor to the Crystal Palace put it, "The blue painted girders . . . range closer and closer together until they are interrupted by a dazzling band of light—the transept—which dissolves into a distant background where all materiality is blended into the atmosphere."[38]

Practices and technologies of envelopment that gave shape to the balloon as a surface of captivation also provided opportunities for engineering structures defined by distinctive experiences of being inside. The inflatable dome exemplified this. From the early decades of the twentieth century, the dome was proposed and employed in various industrial, military, and aesthetic contexts. During the latter part of World War I, the British engineer and polymath Frederick W. Lanchester proposed using balloon fabric to produce air-supported structures such as portable field hospitals.[39] The practical possibilities of such structures were explored and realized more fully during and after World War II, again stimulated by the requirements of the military. In the Unites States, Walter Bird, like Lanchester, developed air-supported structures as low-cost building solutions for housing radar antennas. The success of these structures encouraged Bird and others to explore using similar structures in a range of industrial and domestic contexts, including, most famously, the mobile theater developed by Victor Lundy for the US Atomic Energy Commission. Numerous other companies designed and exhibited air-filled plastic structures promising mobility, flexibility, and portability in a range of contexts.[40] Advertising for these structures emphasized advances in material technology, particularly the development of new forms of plastic, including Mylar, the invention of which seemed to herald a revolution in inflatable architecture.[41]

The allure of such forms of envelopment also lay in the possibilities they offered for critiquing and reworking the qualities of more traditional architectural envelopes. In particular, artistic experiments with inflatable architecture emphasized its lightness, ephemerality, and mobility, while also foregrounding the accessibility and relative affordability of the materials used in its construction, and the "DIY" quality of its assembly. Inflatable architecture simply seemed more vital than rigid and static architectural forms. As two members of the French Utopie movement, Claude and Léon Gaignebet, wrote in the late 1960s, "It is the balloon which confers its visible form to the breath it contains; but it borrows from that same breath its tension, its appearance of being swollen with vitality. . . . At each instant every non-rigid container exemplifies, in its form, the dialectical relationship between container and contents. On the contrary, the rigid container is indifferent to its state of repletion."[42] Another appealing aspect of these environments was their responsive openness. Rather than being hermetically sealed, they were supported via a continuous supply of low-pressure air from the outside. Architectural critic Georges Teyssot wrote that in these structures "the flexible membrane of air-filled volume acts as the skin of a living apparatus plunged in a given milieu. Such an atmospheric architecture creates climatic islands that operate as controlled, homeostatic systems."[43]

During the 1960s, however, the allure of inflatable architectural envelopes was enhanced further, both by advances in materials, and by increasing interest in experiments that combined this with the use of information technology in order to produce responsive environments. Exemplary here is the 1970 Pepsi Pavilion, at the center of which was a large inflatable dome, designed by figures in EAT (Experiments in Art and Technology), an organization established in 1966 by an engineer at Bell Labs, Billy Klüver, along with Robert Rauschenberg, Fred Waldhauer, and Robert Whitman. EAT's aim was to encourage collaboration between technology, science, and art in ways that were interstitial to disciplinary approaches and conventions.[44] During the late 1960s, it attracted the attention of executives at Pepsi, who, in developing new kinds of atmospheric brand association, were deliberately seeking out artists working in the countercultural circles of New York. What emerged was a project, commissioned by Pepsi, for the construction of an installation at the 1970 Osaka World Exposition, at the center of which was a large balloon dome. The mirror dome on which Klüver and EAT worked for the Osaka Expo continued a tradition of experimenting with architectural forms as part of efforts to engineer the aesthetics of display and consumer experience at expositions and world's fairs. Unlike those in the nineteenth and early twentieth centuries, however, where large tethered balloons featured as platforms for affording visitors a view from above, in Osaka the balloon became the very environmental envelope in which visitors were invited to consume a particular kind of atmospheric experience.[45]

This balloon dome was inflated for the first time in a large airship hanger in Santa Ana, California. Made from aluminized Melinex, a material like Mylar, the dome was designed by Chrysalis, an interdisciplinary architecture research group based at UCLA, which had been experimenting with lightweight structures for playing with experience.[46] Constructed by Raven Industries, the shape of the envelope was maintained by a small differential between the air inside and the atmospheric pressure outside. What was most distinctive about the dome was the mirror effect it generated. Unlike flat mirrors, which produce virtual images—that is, images that appear on the other side of a plane—the mirror dome generated "real-image, three-dimensional, upside-down reflections of audience and performers."[47]

Installed in the Pepsi Pavilion, the visual experience of the mirror dome was complemented by a programmable surround-sound system. In effect, the dome became the center of an environment of immersive media experience, the design of which was informed by theories of cybernetics.[48] As Randall Packer has written, "Klüver's ambition was to create a laboratory environment, encouraging 'live programming' in which artists, performers and engineers collaborated in real-time in front of a live audience."[49] The vision affirmed here was of a nonhierarchical social aesthetics through the collaborative participation of artists and audiences, in which, crucially, visitors were to feel as if they were making choices to which the environment in which they were immersed responded. The point here was not to transmit messages or to generate meaning. Instead, for the artists who participated in the project, the aim was to produce an artwork that served as a captivating milieu in which bodies, information, and atmospheres were continuously mediated by one another through their capacity to sense variations. As David Erdman, Marcelyn Gow, Ulrika Karlsson, and Chris Perry write, "In essence the pavilion acted as a kind of large-scale operating system, or protocomputational environment, into which the expo visitors were invited to co-author and inform the exhibition itself. It performed as an ambient machine, in which hardware is subsumed by the atmospheric effects of software, and materiality de-instantiates itself into responsive networks."[50]

The mirror dome gave spherical shape to the promise of responsive envelopes of experience sustained by combinations of hardware and software, fabric and infrastructure. It exemplified the creation of supple and flexible environments responsive to variations in gaseous atmospheres while also generating atmospheres intended to heighten the affective and responsive capacities of bodies within.[51] This was an experiment in atmospheric envelopment in at least three senses: architectural, informational, and affective. Architectural because it used an envelope engineered to respond to the elemental conditions (in a gaseous sense) both within and outside the Melinex membrane; informational because the responsiveness of the immersive environment depended upon the capacity of technical devices to process and respond to data generated by participants; and affective insofar as the response of the environment to this data was intended to work on the bodies of those enveloped on a level prior to or below conscious attention.

The Pepsi Pavilion, with the mirror dome at its center, was aligned with the wider elaboration of envelopes of consumer experience, and with the generation of alluring atmospheric worlds in which consumer affect was both activated and modulated. To be sure, the engineered fabrication of these worlds has long been part of modern life. However, as cultural historian Fred Turner has argued, the Pepsi Pavilion linked in a particularly innovative way the countercultural tendencies of collaborative experiments between artists and engineers and the ideas, technologies, and practices of military-industrial research and corporate management.[52] Rather than offering a critique of the dominant ideological tendencies in popular culture, Turner argues that the Pepsi Pavilion exemplifies the key elements of a cultural formation dedicated to "soft control, cybernetic agency, and American political hegemony."[53] For Turner, the Pepsi Pavilion was not just about experimenting with new kinds of media experiences. It was also about the engineering of alluring worlds in which inhabitants would become "processors of information, programmed by invisible others."[54] At the center of the Pepsi Pavilion, then, is a vision of soft power, of the establishment of a network of messengers within which experience is enveloped in an atmospheric, dynamic cloud of solicitations to choose and to participate.

TANGIBLE ATMOSPHERICS

The promise of the Pepsi Pavilion was the generation of an envelope of atmospheric immersion that would become an autonomous, self-reproducing, and self-modulating ambient environment in which everything had the capacity to mediate everything else. In some ways, then, the Pepsi Pavilion can be read as an experiment in generating atmospheric allure by acting upon environmental infrastructures of sensing that operate outside and prior to conscious attention. It anticipates the kinds of experiments that have become the focus of recent speculation about how media worlds are becoming ever more engineered through processes of what Mark Hansen calls "feed-forward." While such worlds present the lure of choice and intentionality as incentives for acting, they do so on the basis of an expanded technical infrastructure whose operational present is largely intangible because it takes place prior to conscious awareness.[55]

We need to be careful, however, not to assume that all envelopes of alluring atmospheric experience operate on this basis. Clearly, the elaboration of technical infrastructures of sensing may mean that the conditions which generate atmospheric experience are becoming more and more intangible. By this I mean that they are withdrawing from the grasp of human subjects. However, other experiments provide us with opportunities to think about how the shape of the envelope as a technical and aesthetic process, and the allure of what it presents on its surfaces, provide different possibilities for generating what might be called a kind of tangible atmospherics. By this I mean a form of atmospherics generated through the promise of coming into contact with the surface of an entity or object. This can be glimpsed in the shape of another experiment with which Billy Klüver was involved during his time at Bell Labs. While other engineers were experimenting with new technologies of communication in the shape of satellites, Klüver pursued collaborations between technology, science, and art, and developed working relationships with a number of important artists, including Andy Warhol. According to Klüver, in 1964 he and Warhol had been trying to develop helium-filled floating light bulbs. When it became clear this was unachievable, they began experimenting with other materials, stumbling on a solution by chance. Klüver relates, "A colleague found a material called Scotchpak, which the US Army had used to vacuum-pack sandwiches. The material was relatively impermeable to helium and could be heat-sealed. Andy asked if he could make clouds. While we were figuring out how to heat-seal curves, Andy took the material and simply folded it over and made his silver clouds."[56]

What emerged was *Silver Clouds*, a work first shown in 1966. Warhol described *Silver Clouds* as an effort to go beyond a painterly approach to art by making something that floated. He recounted, "I asked Billy Klüver to help me make a painting that floats, and he thought about it and he came up with the . . . silver . . . and the idea is to fill them with helium and let them out of your window and they'll float away and that's one less object . . . to move around."[57] This vision of a floating work escaping the confines of a frame or architectural housing did not quite materialize in the finished work. Instead, *Silver Clouds* consisted of a large number of metalized pillow-shaped balloons filled with helium and oxygen floating through a

FIGURE 3.2 Andy Warhol, *Silver Clouds*, 1966, at the *Warhol Unlimited* exhibition at the Musée d'Art Moderne de la ville de Paris, December 2015. Photographer unknown. Wikimedia Commons.

given space. Insofar as it is always installed in a particular space, and needs to be contained, *Silver Clouds* is a circumstance-conditioned work: its volume and relational density depend upon the size of the room in which it is installed. But *Silver Clouds* is also air-conditioned: the shape and movement of the balloons depends upon the dynamic relation between different gases, and on the circulation of air in the space where it is installed. As Klüver writes of their initial experiment, "The heat gradient between the floor and the ceiling created a slight pressure differential, and with paper clips as ballast, we balanced them so they would float halfway between the ceiling and the floor."[58]

The appeal of *Silver Clouds* has much to do with its interactivity, and with how the tangibility of the objects of interaction contributes to the alluring intangibility of its atmospheric spacetime. This play between the tangible and intangible was central to the use of Warhol's pillows in the generation of atmospheric spacetimes as part of another, more obviously choreographic work. In 1968, Merce Cunningham incorporated some of the balloons from *Silver Clouds* into a performance piece called *RainForest*. In an interview in 2005, Cunningham outlined how this came to be: "I was with Jasper Johns at an exhibition and Andy's pillows were just piled in a corner. I immediately thought they would be marvelous on stage because they moved, and they were light, and they took light. So I asked Andy and he said, 'Oh sure.'"[59] The qualities of the balloons elicited distinctive ways of moving from the dancers: "You had to push, not kick, to get them to float. When we first did *RainForest* they had only had one rehearsal with the pillows, and a lot went out into the audience." In *RainForest* then, the silver balloons became free, "unmanageable," untethered participants that added a sense of uncontrolled spaciousness to a work of air-conditioned choreography. As Marcia Siegal wrote, "The pillows behaved in their own fashion. They skittered along on end at the slightest disturbance of the air. They drifted about the floor or out into the wings."[60] The untethered quality of the balloons meant that the work did not always have the desired effect. Commenting on a revival of the performance, Siegel continued, "The stage looked crowded and inert. Somehow the pillows weren't light enough; when they got moved by the dancers, they bounced up and landed like wads of Jell-O. Perhaps responding to their resistance, the dancers began kicking them out of the way and deliberately barging into them. The energy needed to dislodge the intractable pillows spilled over into the rest of the dancer's movement.

Things that I remember as deliberate but calm looked forced."[61] For Siegel it was not simply enough to use floating things in order to add a kind of aerial atmospherics to a choreographic work: what mattered was the "density and consistency of the space" created between the pillows and the dancers.[62]

By foregrounding this more obviously artistic work here my aim is not to suggest that it escapes the kind of commodity worlds of the Pepsi Pavilion. Warhol, after all, was masterful at generating value from the symbolic reproduction of the commodity form. Equally, versions of the kind of interactive envelopes above can be and are deployed in a range of contexts to generate and extract value from these contexts. On one level this might simply remind us of the use of balloons to add atmospheric allure to occasions and events of many different kinds. But it also helps us think about how the play between the tangible and intangible can be and is used to add affective density and intensity to highly commodified spacetimes of atmospheric experience. For instance, Tangible Interaction is a Vancouver-based company specializing in creating "sensory installations where people participation is key." Their customers include Samsung, Nokia, Ford, Heineken, and MTV, in addition to bands such as Arcade Fire, Green Day, Coldplay, and the Chemical Brothers. Among the "experiential products" sold by Tangible Interaction, through a subsidiary called Crowd Activation, are Zygote Balls: "large inflatable LED" spheres "that react with coloured light to touch," allowing the user to "bring a crowd together" and transforming a space into "a multi-dimensional, interactive playground."[63]

These zygotes and the multitudes that gather around them are a long way from those that gathered around balloon launches during the late eighteenth century. However, the balloons at the center of these multitudes are linked by their shared status as envelopes that participate in the generation of an atmospherics in which consumer experience, and its value, is intensified. In both contexts, the presence of the balloon reminds us that allure is not simply a metaphysical property that resides in objects but is also a quality of an entity or occasion that is actively fabricated in the play between tangibility and intangibility, and in the tension between what is presented on a surface and what remains withdrawn. And in both contexts, part of the allure of this atmospherics is the presence of a sphere that draws our attention, often expressed through a gaze upward, via its lightness, color, and movement.

Envelopes of atmospheric interactivity can be extended skyward in ways that echo early balloon launches while also employing technologies of connectivity to radically distribute a sense of being involved in the generation of this interactivity. Consider, for instance, the work of architect and artist Usman Haque. His aerial works include *Sky Ear* (2004), in which a "cloud" of one thousand helium balloons was released in nets into the air at the Greenwich Maritime Museum in London. The work is described as a

> non-rigid carbon-fibre "cloud," embedded with one thousand glowing helium balloons and several dozen mobile phones. The balloons contain miniature sensor circuits that respond to electromagnetic fields, particularly those of mobile phones. When activated, the sensor circuits co-ordinate to cause ultra-bright coloured LEDs to illuminate. The 30m cloud glows and flickers brightly as it floats across the sky. As people using phones at ground-level call into the cloud (flying up to 100m above them) they are able to listen to distant natural electromagnetic sounds of the sky (including whistlers and spherics). Their mobile phone calls change the local hertzian topography; these disturbances in the electromagnetic fields inside the cloud alter the glow patterns of that part of the balloon cloud. Feedback within the sensor network creates ripples of light reminiscent of rumbling thunder and flashes of lightning.[64]

These kinds of experiments are becoming increasingly common and complex, as the logics of interactivity come to define what it means to generate notionally public forms of atmospheric envelopment. Just as Pepsi entered into a collaboration with EAT on the mirror dome, contemporary versions of these experiments often also involve forms of experimental association between artists and corporations across different sites and surfaces of captivation. *Skies Painted with Unnumbered Sparks*, for instance, is a work by artist Janet Echelman installed in Vancouver in 2014 to coincide with a TED conference. Emerging through a collaboration with Aaron Koblin, creative director of the Data Arts Team in Google Creative Lab, Echelman's work is a seven-hundred-foot-long net of braided synthetic fiber strung between buildings. Designed to move in response to the wind, the color of the work was also intended to respond to the input of individuals who could connect wirelessly to it through mobile devices. The operation of the work is described thus: "At night the sculpture came to life as visitors were able to choreograph the lighting in real time using physical gestures on their mobile devices. Vivid beams of light were projected across a massive scale as the result of small movements on spectators' phones."[65] As far as the artist is concerned, then, the work is not just a physical artifact but also a kind of captivating surface, responsive to the wind, and onto which is projected a web browser: "The lighting on the sculpture is actually a single full screen Google Chrome window over 10 million pixels in size."[66]

As a device for doing atmospheric things, this work is clearly not the same as a balloon. Nevertheless, it can be grasped as an alluring atmospheric envelope. In this work the surface of the envelope is opened up as a media screen and, crucially, in ways that mean that the encounter with this surface becomes coterminous with the perception of sky as a screen of illumination. Such a work points us to the ongoing elaboration of the forms and technical capacities of what James Ash has called "interface envelopes."[67] For Ash, these envelopes are immersive spacetimes in which different objects are drawn into affective relations. As Ash also notes, the screen, and the touch screen in particular, is increasingly important as a surface of luminous tangibility that provides the locus for what he calls "envelope-power," by which he means the capacity to generate locally alluring spacetimes within which bodies are affectively engaged and immersed. Drawing attention to Echelman's work does not diminish the importance of the screens about which Ash writes; instead, it expands the range of devices and experiments that facilitate new forms of envelope power. This kind of experiment points us to how the very idea of the envelope as a surface of captivation may become more expansive, and in ways that generate new spacetimes of allure. At stake in such experiments is the question of how the allure of atmospheric things is intensified through technologies and practices of envelopment in which the differences between inside and outside, surface and depth, ground and air, are no longer either straightforward or easy to discern, but which nevertheless continue to generate new sources of value. What is at stake is how envelopment mediates the affective intensity of atmospheres, producing, in turn, new experiences of immersion. And what is also at stake is the question of which bodies are exposed to this form of envelope-power, under what conditions, and to what ends.

Because of what is at stake, any claim about the metaphysical importance of allure as a condition for grasping the force of things, objects, and entities needs to be qualified in important ways. First, allure must be understood in relation *both* to those qualities of an entity it presents to other entities *and* to how the essence of that entity is withdrawn from other entities.[68] This means that the complex entangled meanings of allure cannot be ignored, meanings already embedded in a range of practices that precede Graham Harman's deployment of the term. The relation between allure and the form, promise, and enchanting power of commodities and consumer worlds is of particular significance.[69] Indeed, without this qualification, the claim that something of the object is always withdrawn might be seen to echo a form of commodity fetishism; as Patricia Clough notes, there is a risk that it reinforces the idea that something of the entity is always ungraspable, withdrawn from us, beyond critical analysis.[70] However, as Jane Bennett has shown in relation to enchantment, to foreground this aspect of things is to open ourselves up to its capacity to unsettle us, to move us to think in different ways through qualified experiments in thought, affect, and percept.[71] We might think of allure in a similar way: to foreground allure is to attend to how the qualities of particular materials—color, sheen, and feel—contribute to the generation of envelopes of captivation in which to experiment with thinking, feeling, and moving differently.

Second, and equally importantly, it is not only entities or objects that are alluring. The force of the atmospheric, excessive as it is of the category of object or entity, is also alluring, both in the sense that its affective appeal can be actively designed, engineered, choreographed, and in the sense that something of it always remains beyond prehension, sense, or perception. Clearly, atmospheres have distinctive qualities, which may be thermal, sonorous, visual, and so on. However, no matter how completely we feel we are immersed in them, atmospheres never present themselves fully to us, or to any entity for that matter. Equally, no matter how enveloped an entity is, it is never entirely coterminous with an atmosphere. Because of this, the allure of atmospheric things is characterized by an aesthetics of everyday sublimity as much as any entity is. The prospect of atmospheric envelopment is also alluring in other ways. It is alluring in an affective sense because it promises different possibilities of being and becoming enveloped, not all of which are benign. It is alluring in a meteorological sense because it promises different possibilities for the technical sensing of forces and movements that can become actionable and operational as infrastructures of transmission, distribution, and dispersal.

Third, and crucially, the allure of atmospheric things is not simply a given. It does not just reside in entities, already inherent as a constitutive element of their existence. Nor is it abroad in the world, waiting to be sensed. Rather, allure is part of the cofabricated technicity of things as they become atmospheric. The allure of atmospheric things is coengineered in ways that draw together a range of agencies, elements, forces, and bodies in arrangements that are variously familiar and novel, but never reducible to the terrain of the human. Indeed, for atmospheric things as much as for anything else, "the allure in allure is largely produced by the creation of worlds in which boundaries between alive and not alive and material and immaterial" are increasingly blurred.[72]

Given this, in a world where the form of the commodity has become increasingly hazy, it may still remain useful to pay attention to how relatively simple technologies of captivation such as the balloon participate in the engineering of allure in contemporary economies of experience. Indeed, examining this in relation to the generation of affective value in different forms of contemporary capital life could be one way of helping us glimpse the development of new versions of envelope-power even as these versions revise much older tendencies to be drawn into the alluring orbit of things. To suggest this is not however to reduce the process and experience of envelopment to the status of a condition for the circulation and distribution of consumer atmospheres and affects. Envelopment can also be affirmed, albeit cautiously, as a speculative process and technology for experimenting with the allure of atmospheric things in ways that allow us to grasp other forms of ethicopolitical relation that are tensed between entities and the elemental conditions to which they are exposed.

4   RELEASE

In the early 1980s I remember watching, without really understanding, a television miniseries called *The Martian Chronicles*, adapted loosely from a book of the same name by Ray Bradbury.[1] Not quite a novel but much more than a series of isolated stories, Bradbury's book is vaguely episodic without the sense of narrative sequence implied by that term. Two scenes from the TV series linger in my memory. In the first, the owners of a diner, established on Mars in anticipation of the arrival of a large workforce of miners, watch through a telescope as the earth explodes in nuclear Armageddon. As they absorb the impact of what has happened, a voiceover narrates, "A million years in the future, a million light years away, some civilization will perceive a brief flickering in the heavens. Will they know us? That what we had was worth preserving? No. A falling star perhaps. And their telescopes will continue gazing into the universe. But we will be gone."[2]

Something about this scene resonated. It must have been something to do with the atmosphere of the times. Something to do with how, at the age of eight or nine, the prospect of the end of the world was all too real. Something about how, even in Ireland during the early 1980s, the possibility of mutually assured destruction generated a background atmosphere within which childhood dreams of the future came to make sense and were rendered fragile, made to feel palpably finite. Such a sense of finitude was layered with, and complicated by, a sensibility shaped by a vision of the afterlife derived from Irish Catholicism. The possibility of *living on* after the end of the world, after the end of the body, was also real: it had the felt reality of a promise rendered in various representational and figurative forms. Angels existed. Dead relatives had merely gone somewhere else. Thunder was god moving the furniture, and the sound of it, already ominous for a young child, was heard and felt as the expression of something heavenly.

The plausibility of the promise of an afterlife perhaps explains why a second scene seemed to linger in my memory. Before returning to watch it on YouTube, I remembered it as floating blue orbs appearing before and speaking to humans. In fact, the scene features Episcopalian priests who travel from Earth to Mars looking for new sins to absolve. The creatures they find on the planet take the form of luminous bright orbs, beings that have transcended physical form and escaped from sin to live in the wind and sky. When the globes finally speak, they reveal themselves as eternal things, living apart from all other beings, angel-like. The orbs reject the offer by one of the priests to build them a church. Having no need of any housing, any protective envelope, they live their lives through the condition of their exposure to the elements.

In Bradbury's story an encounter with these orbs generates a memory for one of the priests, Father Peregrine—a childhood memory of Fourth of July celebrations, a memory of "dim faces of dear relatives long dead and mantled with moss as Grandfather lit the tiny candle and let the warm air breathe up to form the balloon plumply luminous in his hands." And the precipitation of this memory in turn overwhelms Father Peregrine with the desire to commune with the blue spectral forms in the sky before him: "'I must speak to them.' Father Peregrine rustled forward, not knowing what to say, for what had he ever said to the Fire Balloons of time past except with his mind: you are beautiful, you are beautiful, and that was not enough now. He could only lift his heavy arms and call upward, as he had often wished to call after the enchanted Fire Balloons, 'Hello!' But the fiery spheres only burned like images in a dark mirror. They seemed fixed, gaseous, miraculous, forever."[3]

The story from which the scene with the blue orbs is drawn is called "Fire Balloons." As Bradbury explained in a much later essay, "Fire Balloons" was partly autobiographical, insofar as it was inspired by a very particular memory of his own childhood experience of Fourth of July celebrations. Bradbury, then five, would light and release fire balloons with his grandfather. He describes how the "balloon whispered itself fat with the hot air rising inside." And yet, for all the anticipation of its ascent, Bradbury remembers that he "could not let it go," needing to wait for his grandfather to give him a "gentle nod of his head" before he could "at last let the balloon drift free, up past the porch, illuminating the faces" of his family.[4]

Through their light and heat and ascent, the fire balloons in Bradbury's accounts are envelopes for doing atmospheric things in an affective and meteorological sense: they create the glow that edges the moment of release with the halo of occasion. But for the older Bradbury, and for Father Peregrine, they are also experienced as the conjuration of absent presences, precipitating memories of long-dead relatives. Even for the younger Bradbury, fire balloons are tinged by melancholy affects, with a sense of loss: their release reveals the fact that the palpable atmosphere of an occasion, however lively, is always haunted by its inevitable disappearance, and by what is already gone, no longer present: "Once released, it was yet another year gone from life, another Fourth, another bit of Beauty vanished. And then up, up, still up through the warm summer night constellations, the Fire Balloons had drifted, while red-white-and-blue eyes followed them, wordless, from family porches. Away into deep Illinois country, over night-rivers and sleeping mansions the Fire Balloons dwindled, forever gone."[5]

Bradbury's stories foreground how the simple act of release participates in the modulation of affective spacetimes of experience and memory. Through its luminous ascension into an aerial and heavenly beyond, the balloon, once released, embodies the promise of transcendence, the promise of an incorporeal afterlife that nevertheless remains in the present as a possible object of recognition or as the shape of a feeling of what things might become. And the form of the fire balloon in particular, one that combines fragility, lightness, and luminosity, amplifies the palpability of the promise of this transcendence.

However evocative this account is, such acts of release need some qualification. They are contested and controversial for all kinds of reasons, not least of which are the risks they pose to animal life and property, a fact rendered brilliantly in Elizabeth Bishop's poem "Armadillo," in which she writes of the period, during the summer months, when "illegal fire balloons" are released in Brazil to honor saints, and their "paper chambers flush and fill with light." These balloons drift up, "receding, dwindling, solemnly and steadily forsaking us." But in a downdraft they might suddenly turn "dangerous," destroying the nests, habitats, and lives of animals.[6] In many jurisdictions, including Brazil, the release of balloons, whether fire-fueled or helium-filled, is now either regulated or proscribed on the basis of the various threats it poses. While this fact might certainly make the idea of releasing balloons in this way more difficult to affirm, it also foregrounds the question of how to account for the ongoing allure of this act. In many instances these acts are celebratory, or, as the popularity of charity balloon racing reveals, promoted as fundraising events. However, as Bradbury's story also shows, the affective atmospheres of this act, and of its memory, are not unidirectional: certainly, they are rather more complex than the emotional vectors of ascension and transcendence would suggest. These atmospheres are about the waxing and waning of affect, about weight and lightness, and about the fraught relation between presence and absence. They are also sometimes about loss and its ongoing weight, about grief for the disappearance of time, bodies, and persons. The occasional atmospheres that gather round the promise of release can, in other words, provide opportunities for thinking about the relation between envelopment and a kind of exposure to the elemental condition of love: about exposure to the otherness of the world, to the condition of finitude, and to whatever is excessive of this finitude. In drawing our attention to this condition of exposure, the simple act of release, even when only imaginative, is an occasion for speculating about the im/possibility of letting go of something that cannot be held as such.[7]

GEOGRAPHIES OF LOVE

In a remarkable piece of writing, the cultural geographer John Wylie explores the complex intertwining of absence, memory, and love through a meditation on a deceptively simple object: the memorial bench. Taken at face value, this object foregrounds the imperative to keep memory alive, to keep a sense of someone going after their passing. Wylie notes a parallel here with much of the memory work in the social sciences and humanities, in which the affective significance of material remains is often understood in terms of how they bring something to presence: working with these remains is therefore often about generating, through careful and sometimes performative styles of excavation and animation, a sense of the lives that lie behind artifact and object, a sense of their hidden depths. Wylie tempers the imperative that informs and animates this work. For him, what emerges from the process of contemplating these benches, or, indeed, from their use, is not so much a sense of spectral presences. Rather, the more he thinks about it, any straightforward connection between the benches and the felt sense of a presence becomes, for Wylie, increasingly difficult to sustain. As he puts it, "More than anything else, it seemed to me, the benches had an untethered quality, free-floating; there was something oneiric and aerial about their situation high and lost on the hillside, pitched open to the elements. And in this way they in turn demanded something lighter, a drifting account, a passing-through and a passing-over."[8] Going further with this ungrounding, he observes the double sense of lightness and evanescence characteristic of that which is lost, and, equally, of the effort to grasp this through a philosophically informed thinking. The loss "articulated" by the practice of memorializing via the benches can only ever be grasped as a "sense of something slipping away and being carried beyond." This sense, continues Wylie, can be likened to the difficulty of trying to catch something being blown by the wind. Even when caught, what remains is the feeling of "something essentially lost and out of reach."[9]

Wylie accentuates the lightness of something always being lost in the midst of even the most concrete of things in order to foreground and problematize the tensed relations between presence and absence. And he does so as part of the work of apprehending what he calls the "geographies of love." For Wylie, to speak of the geographies of love is not only to name a project that involves the rendering present of something or someone now departed.[10] It also involves grasping our radical openness to the world as central to the possibility and condition of love: a kind of constitutive absence, always in excess of efforts to bring it to book or account. Rather than delineating the spacetimes of gathering, of fusion, or of union, Wylie reminds us that the "geographies of love might instead describe a separation or rupture" that characterizes a more general "*exposure* to the other."[11] By thinking through an encounter with the bench, Wylie shows how being in love might be about the tensed space between envelopment and exposure. His argument exemplifies further how thinking through atmospheric things offers occasions for apprehending the elemental qualities of spacetimes while also pointing to the limits of such apprehension.[12] He shows how objects can become things foregrounding whatever escapes or evades their form. And one of the interesting aspects of this is how Wylie frames the benches in terms not of solidity but of lightness. In doing so he avoids any neat association between physical weight and a sense of levity, between being grounded and being ungrounded.

The material properties, the affective capacities, the organized form, and the durational specificity of things make a difference to their participation in the geographies of love. So, where Wylie opens up and sensitizes us to the possibilities and difficulties of writing these geographies, it might be possible to push this further by foregrounding how the specific properties of things shape these geographies in different ways. This would involve attending to how the relations we have with different things are not the same, and furthermore, how this difference can be sensed in different ways. This is not to try to parse or classify love, nor to differentiate it with respect, for instance, to hate. Rather, it is to trace how things participate in the spacing of love as that which, following Jean-Luc Nancy, can be taken to be the condition of an opening onto the world: love is the condition of being exposed to, and becoming vulnerable to, the affective force of this openness.[13] This condition makes its presence felt in different ways: it shows up in, gathers, shapes, and rends the affective life of bodies in different ways and to different degrees of intensity. It is often felt as a vague variation in an atmospheric milieu that never crystallizes as an entity. And it is often disclosed via encounters with and stories around and about simple devices for doing atmospheric things.

GRIEF AND THE AFFECTIVE LIFE OF THINGS

The opening or exposure of which love is characterized is felt in the possibility and experience of grief. In a number of publications, Judith Butler has outlined how the possibility of being grieved is a necessary condition for life to be valued, and loved as such. In *Precarious Life*, Butler writes, "What grief displays is the thrall in which our relations with others hold us."[14] For Butler, grief is therefore experienced not only as the loss of the other but also as the loss of the sense of self dependent upon the other. However, and beyond this, grief discloses the loss of something else, the loss of something for which there is "no ready vocabulary": a kind of "relationality that is composed neither exclusively of myself nor you."[15] As Butler makes so clear, part of what makes grief so difficult to endure, then, is the fact that at the center of loss is the sense that it is not possible to know what has really been lost.

One way to grasp the vague relationality of which Butler writes is in terms of an atmospheric spacetime. Grief, in these terms, is a disquieting, vague modification of the feeling of that spacetime. Relatedly, Dana Luciano has written of grief as "the corporeal response to the affective residue of the vanished past in the present tense."[16] Grief, she suggests, alters the body's role as a kind of spatiotemporal reference point, modifying profoundly its orientation to pastness and futurity. The grieving body falls out of time insofar as "everything but the past fades away and yet, at the same time, remains."[17] Even then, however, grief is not primarily a process of individualization, nor a collapse into interiority. As a disturbance of and within bodies, grief is always also experienced in relation to wider patterns of practice, norms, and value. The felt spatiotemporality of grief punctuates the rhythms and routines of everyday life while also being shaped in relation to the force of sedimented expectations about conduct and comportment, expectations that, of course, vary enormously. This also means that grief can be and is political: it is always subject to a series of demands and expectations about the proper articulation of the self in relation to the contingency of life. Equally, grief can become part of the elaboration of powerful affective complexes that work to foreclose the possibility of critique, dissent, and disagreement. That is not to say that grief is always antipolitical because it closes down disagreement. Instead, as Butler has argued, it might be possible to use "grief as a resource for politics."[18]

Important as her discussion is in relation to the significance of grievability as a political horizon, Butler tells us little about how the spacetimes of grief are both shaped and in some sense made explicit in relation to the presence of different kinds of artifacts and things. To foreground this is to think about how the relations displayed by grief are not only human, nor only social in the sense in which that term refers to associations between humans: these relations also involve the presence and absence of nonhuman things in different forms of affective life. In relation to grief, of course, there are different ways in which these relations can take shape and be traced. The first, and most obvious, is through attention to the material cultures associated with death and dying, and, in particular, various forms of memorialization.[19] Relatedly, if also more obviously politically, a second way is through the relation between grief and sites associated with particularly traumatic events. A third way in which this relation can be traced is through how artefacts associated with a person act as reminders of the presence and absence of that person. On one level this is about how, once the body is gone, artefacts remain as what performance theorist

Peggy Phelan calls "mute witnesses" to the life of this body.[20] But it is also about the fraught relation between presence, absence, and the process of letting go both of artefacts and the lives in which they were implicated.

ACTS OF RELEASE

Sara Ahmed reminds us that "happy objects" can also be things whose presence is shadowed by the loss attendant upon grief.[21] Frequently understood as a "happy object," one associated with festivity, levity, and fun, the balloon, through the act of release, is also often at the center of occasions of profound sadness.[22] Indeed, notwithstanding its prohibition in many jurisdictions, balloon release remains one of the more common practices used in popular acts of memorialization to mark the death of a loved one or the anniversary of this death. A balloon is not a bench, however, and offers a different set of possibilities for practices of memorialization. Much more so than benches, the form and capacities of the balloon amplify the sense of lightness, absence, and loss evoked so beautifully by Wylie. Moreover, the temporality of the balloon as a shaped form differs from that of the bench. The bench is a gesture toward persistence, while balloon release gestures toward something more fleeting, impermanent: nothing much of the object is supposed to remain after its release. In theory at least, the balloon, once released, is often never intended to persist as an artifact or memento, and the spacetimes of experience it gathers are less obviously site-specific and more atmospheric than those around a bench. And, of course, balloon release embodies perfectly the promise of a departure that is also a form of transcendence, a form of going home, of a release from the earth, a promise that remains so central to the material cultures and affective geographies of death and dying, and of mourning and grieving.

Used in events of memorialization for all people of all ages, there is an especially strong affinity between balloon release and the death of children. The information on the website of the UK-based Balloons for U is typical of how the significance of balloons in this context is understood:

> Balloons released at a graveside or outside a crematorium are a growing tribute to conclude a funeral service. It helps give comfort, and depending upon your faith, helps to symbolise the soul or spirit rising towards the sky and the letting go of the physical presence of a relative or friend. Releasing the balloons and watching them ascend is a very poignant and solemn act of saying goodbye to a loved one.

FIGURE 4.1 Release of thirty-two balloons in September 2007 before a football game with East Carolina to commemorate victims of the Virginia Tech shooting in April 2007. Photographer unknown. Wikimedia Commons.

This is particularly memorable for children at a funeral. Relatives can write messages onto paper notes organized and collected up a day before the service which are then rolled up and inserted in the balloons before they are inflated. Or you can simply write a message onto the biodegradable helium-filled balloons on the day and release one by one as you say good-bye.[23]

Such acts of balloon release have also become part of semiformalized and at times officially recognized events through which life and death become politicized in different ways. In 1988, then president Ronald Reagan designated the month of October as Pregnancy and Infant Loss Awareness Month in the United States. In proclaiming the month thus, Reagan noted the "great tragedy involved in the deaths of unborn and newborn babies," and called "upon the people of the United States to observe this month with appropriate programs, ceremonies, and activities."[24] In 2002, October 15 was designated in the United States as Pregnancy and Infant Loss Remembrance Day, and it continues to be marked through a range of practices, including balloon release. For instance, a Christian organization called Hope Mommies encourages balloon release on that day, inviting people to share images of their own balloon releases or to sponsor a balloon released by the organization with the name of a baby. The overall aim is to "flood the skies on October 15 with a myriad of colors, and encourage hope in families by saying, 'you are not alone. We remember your babies.'"[25] Other support organizations and groups encourage similar rituals. For instance, in publicizing their Third Annual Balloon Release in Mayor's Park, Visalia, California, on October 15, 2013, an organization called A Legacy of Love used a photo of pink and blue balloons rising into the sky with notes attached, accompanied with the tagline: "For some moments in life there are no words."[26] There are myriad similar events that take place at a range of scales, and for a variety of purposes.

The popularity of this practice is not only due to how it symbolizes transcendence. Collectively, the presence of balloons gives color and density to the atmosphere of a shared occasion that is already composed of many different affective possibilities, and balloons are atmospheric envelopes around which gather a cloud of affects: hope, grief, joy, happiness. Holding on to the balloon functions to generate a sense of the presence or memory of the person to whom the event is nominally dedicated. In turn, letting go can be understood as a kind of gift to the deceased person, especially when a message is attached to or written on the balloon. Equally, release can and has been described as akin to the letting go of grief. Indeed, such is the affective resonance of balloon release in popular culture that it is frequently used in books written for those coping with grief. A search on Amazon offers up *Balloons for Mary: A Children's Book about Grief and Coping with Death*; *The Blue Balloon: Journey through Grief*; *Ella and the Balloons in the Sky*; *Balloons on the Mailbox: One Mother's Heartbreaking Story of Her Daughter's Death*; *Purple Balloons for Daddy: A Child's Journey through Grief*; *Balloons to Pawpaw: Letting Go of the Pain while Holding On to the Memories*; *Balloons for Grandpa*; *White Balloons: A Memoir*; *A Bunch of Balloons: A Book—Workbook for Grief.*[27]

BEING FOUND

Something about the act of release seems to make a difference. Too many people continue to do it for it to be otherwise. It might be all about the symbolism of the act, about the hope for the transcendence of something more than substantial. Release is, however, always also a kind of multivectored affective event, one that turns now around the assumption that whatever is released might just disappear, and now around the hope that it might show up somewhere, for someone, for something, even if it will never be us. The release of the balloon embodies the promise, however tenuous, that something may be found.

WHEN I WAS YOUNGER, I prayed before I went to bed each evening. I prayed in what, following Patricia Clough, I now grasp as the peculiarly metronomic style of repetitious Catholicism.[28] I prayed for people who had died. And whenever anyone I knew died, I would add them to my list, a list that grew longer, and which came to be ordered in a manner that linked chronology and proximity. It began with my grandparents, and tailed off with neighbors, including a boy next door who had drowned, friends of my parents, people I had never met but for whom I felt some sense of affiliation. Eventually there were simply too many dead people to pray for. But I was sure they were all in heaven. And I was certain that they were all in some way together—that they had found each other, that they had realized the dream of the ultimate congregation of humanity.

I don't believe this now. But I sometimes feel myself missing the promise that underpinned it. I missed it when you asked the question, one evening, after bedtime stories, about what happens when we die.

Who find us?

The void opened up by that question was too much for me. The exposure was too much. I said that someone would find us. We would be found.

IN OCTOBER 2012, Reiner Gumprich, aged sixty-eight, found a balloon with a note attached while he was out picking mushrooms in Westerkappeln, Germany. The message read, "In support of Karina Menzies. You will be missed, you were such an amazing person. RIP. Lianey Niki & Megan & Cerys." Karina Menzies, a thirty-one-year-old mother of three, had been killed a few days earlier in an alleged hit-and-run, in Ely, Wales. The balloon bearing the message had been released in Cardiff.[29]

A LIFE

Perhaps because I have children I get to watch quite a lot of animated films. I often get to watch the same film on more than one occasion. Some films don't bear up to such repeated viewing. But a few do. One of these is *Up* (2009), a Disney-Pixar film. The plot of this film may be familiar. Carl is a balloon seller whose wife, Ellie, passes away after they have shared a long and childless marriage. Remembering the dreams of travel they shared in the early days of their relationship, and facing eviction from a home now surrounded by a large urban development project, early on in the film Carl concocts a plan: he will travel to Paradise Falls, a fictional waterfall in South America, based loosely on Angel Falls in Venezuela. He attaches thousands of helium-filled balloons to his house, constructing a device for traveling that carries Carl and an unintended passenger—a young, over-eager Boy Scout named Russell—almost all the way to Paradise Falls. Before the denouement, Carl and Russell have various encounters with a rare bird, which they name Kevin, and a long-lost explorer from the golden age of airship travel who has become obsessed with finding proof of the existence of this bird and feels aggrieved that Carl and Russell might prevent him from doing so.

There are various ways in which *Up* can and has been read critically.[30] Perhaps the most obvious, and the most problematic, is the colonial fantasy it enacts, one that has echoes of Jules Verne's *Five Weeks in a Balloon*.[31] The land to and over which Carl travels is entirely depopulated of humans, inhabited only by animate beings of varying degrees of articulacy: talking dogs and screeching birds. This land becomes a lost world onto which a personal narrative of redemption and recovery can be projected. Relatedly, through the voyage undertaken by Carl, the film enacts a fantasy of escapism. Thus, for Carl, the only way of avoiding the real difficulties faced and endured by people living alone in later life in the shadow of grief is a form of childish escapism. Equally, by using the balloons to transform his house into a flying machine, Carl offers a dream of escape from the political economy of urban transformation and change: after years holding out against this transformation, he appears to give up the difficult struggle over real space in favor of a fantastical and implausible journey. Beyond this, *Up* is also a story that rehearses narratives of the balloon as a device for a kind of masculinist mode of traveling, wandering, and adventure. In the film, travel and its promise become substitutes for a lack of children: after the realization that they cannot have children, Carl hands Ellie a scrapbook with "My Adventure Book" printed on the outside. But they never get to travel. Only Carl does. Seen in this way, *Up* is a boy's adventure story, in which the only real travelers and protagonists are male. And yet these rather obvious elements of the narrative should not blind us to what is also a key aspect of the story: *Up* is a film in which the movement and modulation of affects and the properties of a distinctive device are intertwined in a particularly atmospheric way. We might say, of course, that the staging and deployment of this intertwining is part of all good filming. But *Up* uses the properties and capacities of the balloon in a particularly interesting way in order to explore the relation between envelopment and the elemental condition of love as an exposure to grief.

The film begins with a sequence in which the balloon figures as a childlike object whose allure draws in the two key characters. What follows immediately, and for most adult viewers surprisingly, is another film: a kind of film within a film, in which the arc of the shared life of Carl and Ellie is shown over the course of a wordless four and half minutes. At one point in this sequence, Ellie and Carl are having a picnic, and are looking up at the clouds in the sky. One of these clouds forms into a baby, followed by another. We then see Carl and Ellie decorating a nursery before a change in the tone of the music accompanies a change of scene: Ellie—with her head in her hands—and Carl are in a doctor's consulting room. In the next scene Ellie is sitting in the garden with her back to camera and to Carl. Carl then presents Ellie with the adventure book. They plan their journey, but life always seems to intervene. Many years later Carl finally remembers the dream, and, noticing how old they have become, buys tickets to Paradise Falls. Before he can present them to Ellie, she falls ill. In her last scene, we see Ellie in a hospital bed, and a balloon floats over to her from Carl. Ellie then gives Carl the adventure book, which, unbeknownst to him is filled with pages of their shared daily life as an adventure. As they kiss, the balloon remains above Ellie's head. The camera cuts to a picture of Carl alone in a chapel holding that same balloon.

For many viewers, although perhaps not for my children, the remainder of *Up* is shadowed by the short sequence near its beginning. It is a reminder, as Gilles Deleuze has written, that a life is composed of virtuals, of potentials.[32] And yet, as Deleuze's own writing also reminds us, this claim is no guarantee of anything in particular. It is no promise of plenitude, only that something might happen. So, what shadows the remainder of the film is the sense of the loss of this potential. The remainder of the film, in other words, is not only about the wanderings of a windblown festive object but also about how this wandering discloses the geographies of love as the exposure to the elemental, atmospheric condition of grief. For Carl, of course, there is the grief associated with the loss of Ellie. And the balloon he holds in the chapel might be taken to represent this grief. Equally, the journey of the balloons he tethers to the house and then with which he travels (almost) all the way to Paradise Falls charts the variation in the weight and intensity of this grief. As the helium escapes, the balloon becomes more and more difficult to move, with Carl resorting to dragging it along the ground. In order to keep it mobile and in the air, Carl is forced to jettison some of the material artifacts from the secure and comfortable envelope he has shared with Ellie. And, toward the climax of the film, Carl is confronted with the difficult decision of letting go of the house completely. He lets the house and the balloons fall into the mist, after which, unbeknownst to Carl, the house comes to rest and settles at a site near the falls to which Ellie had dreamed of traveling.

Central to the affective dilemma of the key character in *Up* is the tension between holding on and letting go, between the lightness and weight of being and memory. Admittedly, in *Up*, the tension seems to be resolved in a very familiar, and very redemptive, way. In *Up*, as in many other scenes of popular culture, grief is—must be—made to work: that is, it must be endured as part of a process of self-amelioration. Following Lauren Berlant we might think of Carl's journey as a "scene of sentimental education." In such a scene, "death must be meaningful, engendering knowledge that, in moving us beyond the finality of another ending, performs and confirms a future in which we are not abandoned to the beyond or the beneath of history."[33] And yet, at the same time, the film also remains haunted by and never lets go of a particular kind of grieving of the loss of a life, a life that never was but could have been. This is an objectless sense of grief. A grief with a minimal hinterland of artifactual reminders. Despite the fact that early on in the film we see Carl and Ellie painting a nursery, there are no objects or things to which this kind of grief for lost potentiality can be attached. This kind of grief is not, indeed cannot, really be addressed in the film. Clearly, it is addressed rather obliquely: the adventure book given by Ellie to Carl appears to document a life that has been lived. But there are no tokens or mementoes that perform the work of memory in relation to the loss of something that has never really crossed a threshold of public grievability around which an atmosphere can gather. There is nothing of which to let go. No way in which either objects, or that which they might bear witness to, might be expunged. At the center of the film, therefore, is a grief for the absence of a life that could have been properly grieved.

GRIEF WORK

Julian Barnes's *Levels of Life* (2013) is in some ways remarkably similar to *Up*. In this short book, Barnes uses the balloon as a narrative device to move between the geographies of love and grief. The book is in three parts. In the first part, "The Sin of Height," Barnes provides a thumbnail account of the emergence of the balloon as a wayward technology of travel, the ascent and journeying of which give rise to certain kinds of "moral feeling" and which render the "sin of height" redundant.[34] Through the figure of Gaspard-Félix Tournachon, or Nadar, moments from the history of the emergence of the balloon are layered with observations about the visual possibilities associated with this device. In the second part, "On the Level," Barnes begins to draw together the story of the balloon and a series of love stories. For Barnes, the first flight of the balloon, like the rush of love, is an experiment. To fall in love is to rise into the sky and become untethered with the risk of crash: "We may find ourselves bouncing across the ground with leg-fracturing force, dragged towards some foreign railway line." Because of this, continues Barnes, love is defined from the very outset by the possibility, and in a sense the inevitability, of grievability: "Every love story is a potential grief story. If not at first, then later. If not for one, then for the other. Sometimes, for both."[35]

The shape and form of the balloon largely disappears from the third part, "Loss of Depth," in which Barnes documents his own love/grief story that turns around this response to the death of his partner of thirty years,

Pat Kavanagh, a mere thirty-seven days after her diagnosis of cancer. For Barnes, the inevitability of grief is matched by its unimaginability: it arrives as a shock to life and thought for which there is no adequate preparation. And when it arrives, it settles and envelops him as an affective complex, a differentiated atmosphere of sorts, composed of elemental variations that precipitate as feelings of anger, pain, indifference, and envy. Grief, for Barnes, is a strange kind of affective turbulence, reconfiguring "time, its length, its texture, its function," while forcing the individual to enter a "new geography, mapped by a new cartography," a geography in which the pain of feeling is the only determinant of hierarchy.[36] This part of the book is the document of an effort to navigate this geography and the kind of grief work it involves. This is sometimes "passive, a waiting for time and pain to disappear; sometimes active, a conscious attention to death and loss and the loved one."[37] Barnes is very clear that his grief is not objectless—it is not a longing for the unspecifiable, a longing for something about which we know nothing. Barnes dallies, however, with another loss that makes its presence felt in the new geography of his grief work. It is the loss of stories and figures that might have provided another, more familiar, map with which to make sense of grief: stories of god, stories of the afterlife. Barnes, in a way, toys with the mourning of the loss of these reference points, of these patterns of possible orientation. But, in the end, he does not embrace or turn to these stories and the promises upon which they are based.

This has a bearing on how the balloon figures in the story. In the context of a realization of the absence of a promise of an afterlife the balloon reemerges as a significant presence in the third part of the book. The balloon becomes a device for doing Barnes's grief work. But it does so only as a wayward, windblown entity lacking in dirigibility, a moving object whose course is never willed and that can never really be tethered. In that respect, there is no placing, or retethering, of self or memory at the end of Barnes's book. Nor does grief work become a process of finding a direction, a path, a trajectory. Instead, it becomes about holding on to something that also allows something else to be let go. Barnes ends his story thus, with the following claim: "We did not make the clouds come in the first place, and have no power to disperse them. All that has happened is that from somewhere—or nowhere—an unexpected breeze has sprung up, and we are in movement again. But where are we being taken? To Essex? The German Ocean? Or, if that wind is a northerly, then perhaps, with luck, to France."[38]

Ian McEwan's novel *Enduring Love* (1997) begins in the rolling countryside of Oxfordshire, where a middle-class couple is picnicking:

> The beginning is simple to mark. We were in sunlight under a turkey oak, partly protected from a strong, gusty wind. I was kneeling on the grass with a corkscrew in my hand, and Clarissa was passing me the bottle—a 1987 Daumas Gassac. This was the moment, this was the pinprick on the time map: I was stretching out my hand, and as the cool neck and the black foil touched my palm, we heard a man's shout. We turned to look across the field and saw the danger. Next thing, I was running towards it. The transformation was absolute.[39]

The "it" in this instance is nominally a balloon: a large, unruly, wind-blown entity that arrives silently, surprisingly, to disrupt the idyllic moment. A man, presumably the pilot, is trying to regain control of the balloon, in the basket of which is a young boy. What ensues is an urgent rush to hold the balloon and its passenger down, toward which various individuals, all men, are drawn. Just when they think they have the balloon under their control, however, they are surprised by a gust of wind, the force of which causes the balloon to rise again. What follows is a moment of being suspended, between "one or two ungrounded seconds" and "ruthless gravity," before all but one of the men lets go.[40] Those who remain on the ground watch this man clinging to the rope, dangling for a few moments, before falling to his death.

In some ways, the opening scenes of the novel echo in dramatic fashion the question that functions as a refrain in Barnes's *Levels of Life*: "You put two people together who have not been put together before. Sometimes it is like that first attempt to harness a fire balloon: do you prefer crash and burn, or burn and crash?"[41] After such a dramatic opening, the remainder of *Enduring Love* feels like a slow, gradual deflation, as the characters come to terms with the afteraffects of the event in which they have participated. The philosopher James Williams has written that McEwan's novels are all about events, about how events unhinge us, in the wake of which, as the central character Joe Rose, notes, "transformation" is "absolute."[42] Like any event, it is clear that this is not one whose coordinates can be mapped with any precision. There is a temptation to see the space of this event in Euclidean terms, as if viewed from above: perhaps, like Nadar, from the relative stability of a tethered balloon. As the narrator, himself a scientist

FIGURE 4.2 Still from *Enduring Love*, directed by Roger Michell (United Kingdom, Pathé Pictures International, 2004). Adapted from the novel of the same name by Ian McEwan.

who puts his faith in rationality, describes it, from this perspective "the convergence of six figures in a flat green space has a comforting geometry from the buzzard's perspective, the knowable, limited plane of the snooker table. The initial conditions, the force and the direction of the force, define all the consequent pathways, all the angles of collision and return, and the glow of the overhead light bathes the field, the baize and all its moving bodies in reassuring clarity."[43] But even the narrator recognizes that the geometry of this event or encounter is not easily mapped. This geometry is also topological, drawing together trajectories and lives that had previously been distant, and generating a new multiplicity. The balloon is at the center of an event, a "catastrophe, which itself was a kind of furnace in whose heat identities and fates would buckle into new shapes."[44] This is an event in the sense that Gilles Deleuze and Félix Guattari would describe it: it is a kind of incorporeal transformation, something both beyond and within the bodies of which it is composed.[45]

Enduring Love, then, is about the event of love as an incorporeal transformation, and about the relative speeds and trajectories of this transformation in relation to the bodies caught up in it. It is also about what, following thinkers like Derrida and Serres, we might call the circumstantial qualities of this event and its trajectories.[46] Thus, we discover that the man who falls and dies, Joe Logan, is suspected by his wife—wrongly as it turns out—of having an affair with a student at the time of the accident. For Jed Parry, another of the men who tries to hold the balloon down, the unhinging that ensues is quite literal. He develops a dangerous obsession with Joe Rose—an obsession that complicates Rose's relationship with Clarissa, his partner. This is also a relationship that, in the wake of the event, becomes unhinged, or untethered. Significantly, as with Carl and Ellie, Joe and Clarissa do not have, and apparently cannot have, children. While they seem to have reached an accommodation, with Clarissa having "buried the sadness," the narrator says that "occasionally something happened to stir the old sense of loss."[47] What was revealed on such occasions was "Clarissa's own mourning for a phantom child, willed into half-being by frustrated love."[48] For the most part, however, the "unconceived child barely stirred before the moment passed."[49] The balloon event changes this. According to the narrator, in John Logan and his sacrifice, Clarissa saw a "man prepared to die to prevent the kind of loss she felt herself to have sustained."[50]

What then to make of how McEwan uses the balloon in all of this? The balloon is not the event: the event exceeds incorporation in form. And yet the balloon is not a mere object that could be substituted for any other object as part of the telling of this story. The event would not, could not have taken place without the balloon and its particular properties, its capacity to be both light and terrifying. The balloon in *Enduring Love* is one that deliberately complicates any sense of a straightforward relation between gravity and levity. In the opening scenes it becomes present as an attractive force, a "colossus in the centre of the field that drew us in with the power of a terrible ratio that set fabulous magnitude against the puny human distress at its base."[51] At the same time the balloon is an elemental entity defined by lightness. This is a balloon filled with "helium, that elemental gas forged from hydrogen in the nuclear furnace of the stars, first step along the way in the generation of multiplicity and variety of matter in the universe, including ourselves and all our thoughts."[52] And the fact of its being filled with helium—as opposed to hot air—makes a material difference to the turn of events. Even when tethered, the balloon has enough buoyancy to lift off the ground again when blown by a gust of wind, a gust that we might understand, following Serres, as a kind of circumstantial turbulence.[53] The balloon, in this context, is a device whose specific affective qualities matter even if its essence is to be withdrawn from us, to be always lost.

In *Up*, *Levels of Life*, and *Enduring Love*, the possibility and difficulty of negotiating the tension between holding on and letting go is crucial to the way in which love stories and grief stories develop. In *Up*, a certain untethering becomes the initial event through which Carl begins the process of releasing the weight of the grief he carries for Ellie. He finally lets go, of course, but not into the sky. His grief is released only when, in the form of the house, it comes to be placed, or enveloped, in the shape of the house. He becomes grounded again. In *Levels of Life*, Barnes avoids this resolution, this kind of reassuring emplacement. We are no more in control of the clouds than we are of grief. We don't decide when it arrives, settles, or departs. All we can do is go with it, let ourselves be taken by it in the hope that wherever we end up might turn out to be relatively clement. In *Enduring Love*, the question of holding on and letting go is more obviously framed as an ethical question about how one should respond to the problem faced by another. The other here is not just something that takes form as a person. It is Clarissa's disappointment with Joe's failure to keep holding on to the child who is ascending into the air that unsettles her relation both with Joe and with the impossible memory of the life of her own unconceived child.

In thinking with these stories here, and particularly with stories of childlessness in *Up* and *Enduring Love*, there is a risk that both their authors and I rehearse and reaffirm a particular view of envelopment, one critiqued by Luce Irigaray, of the woman as a kind of maternal envelope, a view in which childbirth becomes a kind of feminine telos.[54] But the risk is worth taking if it allows a way of thinking about how the condition of being an elemental body involves an ongoing exposure to the tension between holding on and letting go, a tension which, of course, remains unresolvable. It is akin to the difficulty Wylie identifies with respect to the geographies of love: the difficulty of trying and always failing to catch whatever spectral presence whose absence is palpable; the difficulty of chasing something as it is being blown by the wind; the difficulty of grasping or losing a sense of loss; the difficulty of sensing atmospheres as they issue from the form of things. There is something similar at work in Butler's discussion of grief: a sense of loss whose object always remains out of reach. And yet this discussion directs our attention to another difficulty—not so much the difficulty of grasping something that will always be beyond us but the difficulty of letting go of something, whether an object, memory, or affect, of whose presence we were never certain in the first instance. The release of the balloon takes places as an opening onto that which can never be grasped. At the same time this release is an act, however brief, of letting go of the weight of the affect of this exposure. It is a lightening of something. Put another way, release is a modification of a relation to something beyond an entity to which we are in thrall and to which grief is an exposure. Or, more precisely, release modifies the capacity of grief to capture these relations.

This might be a way of thinking about what happens during many acts of balloon release and about their ongoing allure. But it might also be a way of thinking about how we understand the wider philosophical problems of our relation with things. Do we, or can we, mourn the passing of the balloon, or grieve for its departure? I am sure my children have done at various points. Like Pascal in Albert Lamorisse's film *The Red Balloon*, they react immediately, sometimes tearfully, to the loss of a companion thing. But this mourning at the coming to an end of an "it" is not—*cannot*—be the same kind of grief experienced at the loss of a "him," a "her," a "you." This might seem an obvious point, but, as Paul Harrison has so clearly articulated, it is at the heart of the kind of affective-ethical implications that emerge in the wake of turns both to a kind of a distributed materialism and to a speculative realism. As he writes, in the wake of these turns, "a concern for the singularity of the other and in particular for the fact that one day this will go, that he is gone, seems more than a little anachronistic."[55] Despite this, Harrison, like Wylie, wishes to retain a form of minimal humanism in the form of the name. The upshot of this is the affirmation of an offering that takes a particular form of address: a prayer. Harrison's treatment of prayer is important. As he notes, prayer is a form of "useless speech," often "held out to an invisible, potentially indeterminate and non-existent addressee."[56] Prayer, in this sense, is not necessarily spiritual but is one way in which "corporeal life turns towards the incorporeal," in an admittedly "gratuitous way."[57] In making this argument, Harrison returns to the kinds of geographies of love to which Wylie draws our attention. As Harrison notes, such geographies are not just about forms of togetherness, proximity, and intimacy; nevertheless, a focus on the occasion, "on the gathered or assembled," can perhaps make us "forget loss itself."[58]

Clearly, the presence of the balloon is a reminder of the occasion, indeed part of the generation of the occasion as an atmosphere of shared, sensible relationality. But in standing apart as a temporary part of this occasion, it brings close that which cannot be held, and in doing so reminds us of our exposure to the other and to the loss of another in ourselves in ways that foreground the difference between this other and the thing through which the other is addressed. In this sense, balloon release might be seen as a form of prayer, one framed by religious notions of transcendence, but it need not be framed thus. It is a gratuitous offering, one borne of the effort to generate a relation with some opening beyond relation, an opening that also serves as a reminder of the difference between things that is not itself a thing. It is possible for me to grieve the loss of that toward which the balloon might be moving but can never reach. And while I cannot grieve the balloon in the same way, to reside in the space between the holding on and letting go of the balloon is to be in the atmospheric spacing of elemental exposure to something apart from me, something always withdrawing from me, but something whose absence cannot but be sensed.[59]

5

As a concept for thinking about the affective materiality of spacetimes, atmosphere presents both a promise and problem. It promises a way of grasping the vague yet forceful, insubstantial yet palpable qualities of spacetimes. The problem it presents is that in speaking of atmospheres as spacetimes, we risk turning them into three-dimensional entities that are no longer vague: they become instead contained and calculable as objects. It is here, perhaps, that the tension between the gaseous and the affective senses of the atmospheric becomes especially stark. The physical properties of gaseous atmospheres are measurable and calculable, even if their variations are not always predictable over a long time horizon; however, the allure of atmosphere as a concept for thinking about spacetimes of affective experience is precisely its resistance to calculability and to the kinds of metric abstractions upon which calculation is based.

These differences become especially acute around the issue of volume. Deriving and describing the volumes of gaseous atmospheres is reasonably straightforward: it is central to understanding the physics of atmospheric phenomena in relation to a set of other variables, including temperature, density, and pressure. The same cannot be said of atmospheres in an experiential sense. If these atmospheres are rendered explicit as volumes they become three-dimensional entities that lose the qualities of vagueness that proves so alluring for those interested in thinking and experimenting with these spacetimes. However, some way of grasping the extensive qualities of experiential spacetimes of varying intensity is still necessary. These spacetimes may not be volumetric, but insofar as they are felt in bodies they have a palpable sense of volume.

An important question, therefore, for any attempt to think through the relation between the gaseous and the affective senses of atmosphere is how to move between two senses of volume. The first sense of volume is that which is calculable as an object of recognition; the second is that which resists calculability, which does not become the object of recognition but which can nevertheless be sensed by a body as the force of something excessive of that body. The shape of the balloon arguably gives form to this problem even further. As a gas-filled envelope, its volume can be calculated precisely. Indeed, early experiments with balloons were implicated in the development of scientific understandings of volume as part of efforts to understand the behavior and properties of gases. As a technical object, the balloon became airborne as a volume of enveloped gas whose properties differed from the elemental milieu in which it moved. Experiments undertaken by one of the earliest balloonists, Jacques Charles, contributed to the emergence of new understandings of these properties.[1] However, the balloon launches and ascents undertaken by Charles also generated atmospheric occasions that were both immeasurable and undeniably palpable, that could be sensed in the bodies enveloped by them while remaining beyond the grasp of those bodies.

Atmospheric volume is therefore both engineered and experienced. In 1897 the engineer Salomon August Andrée and two fellow Swedes, Nils Strindberg and Knut Fraenkel, attempted to fly by hydrogen balloon to the North Pole from a base on Danes Island (Danskøya) in the northwest region of Svalbard. The story of the Andrée expedition is a familiar one, well told in many styles and genres, including film.[2] Two films by Swedish director Jan Troell take the expedition as its focus. Among the scenes in the *Flight of the Eagle*, Troell's dramatized feature film about the expedition, are those in which the expedition balloon is inflated inside the wooden house constructed for it on Danes Island.[3] In these scenes, Troell shows us the volume of the balloon taking shape slowly, breathing into

FIGURE 5.1 Salomon August Andrée, played by Max von Sydow, depicted in the inflated envelope of his balloon in advance of its launch in 1897 from a site in Svalbard. Still from *Flight of the Eagle (Ingenjör Andrées luftfärd)*, directed by Jan Troell (Sweden: Bold Productions, 1982).

form with a rhythmic pulse, while around it the members of the expedition watch, monitoring the birth of this new entity. They do so within another volume of sorts, this one architectural: a shed, or house, in which the balloon is tethered until the meteorological conditions are right for launch. In these scenes, then, volume is an obviously technical achievement, defined in terms of the three-dimensional extent and scale of an object or entity.

Another scene in Troell's film complicates this sense of volume, however. In that scene, Andrée, played by Max von Sydow, approaches the inflated balloon, now full of hydrogen, as it is sheltered in the house. He ducks under the opening at the base of the balloon, and, after an audible intake of air, stands up in the envelope. He gazes around the inside of the balloon, up into it, trying to comprehend, perhaps, the qualities of this space: feeling, perhaps, his breath enveloped in a body, enveloped in turn by hydrogen, enveloped by silk, enveloped by a balloon house itself constructed to shelter the balloon from the force of the elements (figure 5.1). Perhaps at this moment, even for Andrée the engineer, the technical process of giving volume to a fabric envelope is transformed into an aesthetic experience of envelopment. Perhaps, at this moment, the calculably extensive becomes palpable as the alluring intensity of feeling of a spacetime whose qualities are excessive of the metric. Perhaps at this moment, then, something of volume remains irreducible to three-dimensional geometry: an aesthetic, affective quality that to be sure is generated by and related to sheer size and scale but is excessive of these terms of measure.

To think and move between the different senses of volume generated by atmospheric envelopment is to hold in tension volume as measure and volume as qualitative intensity. It is to develop a differentiated concept of volume, to think of envelopment as a process tensed between, on the one hand, what Tim Ingold, following Deleuze and Guattari, calls the "striated space of the volumetric," and, on the other, the smooth space of the voluminous, where the former is an index of metric extension and the latter an index of intensive variation.[4] It is to hold open the category of volume as a property of both gaseous and affective atmospheres, of entities and their excess.

ON VOLUMIZING

An interest in what might be called the "volumizing" of spacetimes is one of the more influential trajectories in recent contemporary sociospatial theory. An important source of theoretical inspiration for this are the writings of the German philosopher Peter Sloterdijk, and especially his three-part elaboration of a generalized theory of the spatial forms that sustain and secure modern life.[5] In these works Sloterdijk develops an account of spatiality that runs counter to the figurative and ontological flattening of spacetimes in much social theory by taking into account how air (or the gaseous) in its multiple manifestations is incorporated in the shapes and forms of inhabitable worlds. In contrast to other metaphors, such as the network, which tend to reduce spatiality to the play of relations on and as surfaces, Sloterdijk proposes instead an environmental account of forms of inhabitable life in terms of the emergence and reproduction of three different emblematic figures, or spherical volumes: bubbles, globes, and foams. The bubble is used by Sloterdijk to develop a microspherology through which to conceive of the birth—metaphorically, physically, and psychically—of inhabitable worlds. It is a self-referential, life-sustaining envelope that makes a space tending toward enclosure, whether in the form of the psyche or the womb, a space that nevertheless remains open to the world. The globe, in contrast, is a sphere that provides a common space of imaginative inhabitation. Through the figure of foam, finally, Sloterdijk develops a "polyspherical" conception of the world in terms of associations of relational aggregations between spheres. In these terms, societies are understood as "an aggregate of microspheres of different sizes (couples, houses, enterprises, associations) bound together like bubbles in a foam mountain, slipping over or under each other without truly reaching, nor effectively detachable from one another."[6]

Sloterdijk's spheres of life are fragile but sometimes enduring zones of inhabitation delimited by semipermeable membranes shared by neighboring spheres that also act to insulate spheres to some degree from disruption. Spheres are never entities but forms of semi-insulated envelopment held in place through relations with other spheres. Movement, both generative and disturbing, passes across and through these membranes such that while each sphere might be relatively autonomous, it is never isolated. And in a world of progressively more dense organizational forms, these movements circulate ever more quickly and with greater and greater consequence.[7] Sloterdijk's theory of spheres is therefore best grasped as a kind of ontological and metaphysical formalism, in which a particular process of shaping comes to define the necessary condition for sustainable spaces of inhabitation and relation. As he puts it, "Spheres are immune-systemically effective space creations for ecstatic beings that are operated upon by the outside."[8]

This is a theory, then, that holds in tension a concern with a transversal geometrical and philosophical formalism in the vital volume of the sphere, and a concern with the materiality of diffuse atmospheres as they propagate across and between bodies. Sloterdijk helps us think, therefore, about the formal properties of envelopes as volumes while also encouraging us to be attentive to the milieu in which these volumes are situated and to the relations upon which they are in some sense codependent. At the same time Sloterdijk sensitizes us to the role of gases in shaping and sustaining spheres and volumes of modern life, foregrounding air's implications, through processes of inflation, envelopment, and respiration, in this life. In a more historically specific sense, then, Sloterdijk shows how critical to the volumetric shape and voluminous qualities of inhabitable worlds are technologies and infrastructures that provide for the circulation, distribution, and use of gases of different kinds, and to different ends, as his account of the use of poison gas reveals.[9]

Compelling and suggestive, Sloterdijk's work is not without difficulties, however. Among these is a potential tendency to reify the form of the sphere such that it becomes an almost atemporal platonic and volumetric shape. Yet the allure of the sphere as a way of imagining a form of life does not necessarily work against a processual and atmospheric materialism if tempered by attention to its emergence through processes of envelopment and inflation, and to the kinds of associations and modes of sensing these processes generate. Equally, there are different ways of developing

Sloterdijk's argument. One is by emphasizing how the rendering explicit of the volumetric is linked closely with the question of how spacetimes become the focus of technologies and rationalities of government organized around the imperative to secure the conditions necessary for forms of life.[10] As recent writing by Stuart Elden has demonstrated, insofar as it is a problem for government, space is more than a surface area to be mapped and controlled.[11] Instead, government involves securing the integrity of the volumetric qualities of space; that is, its depth and multidimensionality are conceived and rendered explicit in relation to technologies of measurement. Elden is careful to note that the volumetric is not just defined in terms of the aerial, however. Rather, it is about different configurations of the relation between verticality and horizontality above and below the surface of the earth.

Such claims parallel other work about how atmospheric processes become the focus for a range of political technologies and rationalities of government.[12] Equally, they echo more general critiques of the volumetric as the appropriation and reduction of space. For Henri Lefebvre, the lived, affective spacetimes of everyday life are commodified as volumes. As he puts it, "Today what are bought (and less frequently rented) are *volumes* of space: rooms, floors, flats, apartments, balconies, various facilities." In these terms, verticality, and the architectural abstractions that facilitate it, separates volumes from the land upon which they depend and with which they should be connected. Through processes of architectural abstraction, volumes are reduced to "surfaces, as a heap of 'plans,' without any account being taken of time."[13]

Framed thus by the volumetric, volume becomes the object of critique in ways that make it difficult to affirm. The volumetric, however, is only one way of thinking of volume in relation to the organization and experience of spacetimes.[14] An emphasis on the volumetric can be supplemented with an emphasis on the *voluminous* qualities of volume. I use this term to indicate the intensive spaciousness of atmosphere. The qualities of this spaciousness are not so easily grasped in terms of the volumetric: they are qualities sensed, for instance, as different degrees of lightness, loudness, density, or gravity. To foreground the voluminous is therefore to draw attention to the fact that felt affective intensity is an index of the volume of atmospheres as much as calculable measure.[15] In these terms, the spaciousness of atmosphere becomes what Gilles Deleuze and Felix Guattari suggest is "something less than a volume and more than a surface."[16]

In the opening pages of the first volume of the spheres trilogy, Sloterdijk dwells briefly upon a painting called "Bubbles" (1886) by John Everett Millais, in which a child has just blown a bubble. For Sloterdijk, there is "solidarity between the soap bubble and its blower that excludes the rest of the world."[17] The blower is temporarily "outside-himself" during this journey, coexisting with the bubble "in a field spread out through attentive involvement."[18] Put another way, the child enjoys the process of giving birth to volume, while remaining immersed in an atmospheric medium whose volume is beyond its grasp yet upon whose dynamics it depends. And the presence of this form transforms the affective relations in this medium.

This reading can be stretched to think about the volume of a balloon. When inflated by the breath of a human, the balloon comes to stand outside that individual by generating a relational-affective space between itself and the source of its volume. Admittedly, there is an important difference. The relative transience of the bubble dooms it to perish at the end of a span of attention short enough that the connection between breath and sphere is obvious: the balloon, in contrast, can and does last much longer, becoming a presence with a duration that effectively sets it apart from the source from whence it was issued. Clearly, balloons are not only inflated by human breath but also by other gases, including compressed air, helium, and hydrogen: nevertheless, at its most basic, a balloon—like a bubble—emerges through a process of envelopment enabling the capture and containment of a gas in ways that are generative of different spacetimes and senses of volume.

As a device for experimenting with volume, the balloon is deployed in both technical and aesthetic ways. To illustrate the latter, we might point, for instance, to how between October 1959 and March 1960 the Italian artist Piero Manzoni made *Corpi d'aria*, or *Bodies of Air*, consisting of forty-five pieces, each of which contained the following: a deflated balloon, a box, a tripod base, and a mouthpiece for inflating the balloon. The pieces could be purchased with the balloon deflated or, at an extra cost, inflated with Manzoni's own breath. In a series of related pieces, called *Fiato d'artista*, or *Artist's Breath*, Manzoni inflated red, white, or blue balloons with his own breath before sealing them and attaching them to a wooden base.[19] Manzoni's works take the simple pleasure of breathing and show how the process of giving form to volume can be commodified. These works are also experiments in duration—of the performance and of the objects that participate in this performance.

In such works, the balloon gives form to breathing as the sensuous refrain of the body's volumetric exchange with air.[20] As Sasha Engelmann has argued in a different context, aesthetic works such as this are important demonstrations of the fact that it is through the volume of our lungs, with all their folds, that we partake of the air within which we are immersed.[21] It is through breath that we sense the world giving volume to us. Through breath we also give birth to volume, sometimes enveloped in shaped form. Breathing is also the edge, the limit, however. We are always one breath away from something beyond us. But we are what Timothy Choy calls "breathers" in other ways than those the direct inflation of these works suggest.[22] We are participants in distributed assemblages of inhalation and exhalation through our implication in networks and devices of power consumption of myriad kinds.[23] This is not always easy to see, even if we have various indicators, from stickers reading "Think about climate change" pasted beside light switches, for instance, to apps of various kinds.

It remains difficult to make explicit as matters of concern the volume of bodies of gas not ordinarily perceived in discrete forms as entities. In 2005 the Victoria state government in Australia funded an advertising campaign intended to demonstrate how much greenhouse gas was being released from domestic sources. What became known as the "Black Balloon" campaign featured helium-filled balloons emerging from a variety of domestic appliances and then rising into the air. One ad begins with shots of various domestic appliances. The top pops off a coffee percolator, and a black balloon begins to inflate from within. Something similar happens from a heating vent in the floor, a TV screen on standby, a light left on in the daytime. As the advert unfolds, more and more balloons appear. The doors of appliances swing open by themselves, and clusters of balloons rise to the ceiling before escaping into the atmosphere. The sky fills with these balloons, and they begin to obscure the view of the cityscape.

There is a relentless sense of urgency and agency to the balloons as they make their presence felt. They appear possessed with intention, striving to escape the confines of the domestic appliances and environments from which they are being issued and extruded. The voiceover, which begins just as the first balloon escapes, goes as follows: "Each balloon represents 50 grams of greenhouse gas. You can't see it, but you produce greenhouse gas every time you use energy. The average home produces over 200,000 balloons every year. Save energy and you'll also save money and reduce your im-

FIGURE 5.2 Still from the "Mother and Baby" ad as part of the Black Balloons campaign. Directed by Adrian Bosich, Exit Films, for the Department of Sustainability and Environment, Victoria State Government, Melbourne, 2006.

pact on climate change. You have the power to make a difference."[24] On the face of it, the black balloons represent the weight, in grams, of greenhouse gas. Yet they manage to do this by giving this mass volume: they give form and shape to the volume of a material ordinarily uncontained and invisible.

In another ad (figure 5.2), black balloons escape from domestic appliances and become trapped beneath the ceiling of a kitchen, watched by a baby whose mother remains oblivious to their presence. The ad materializes the relation to which Sloterdijk draws our attention in his discussion of the child blowing a bubble. If the painting by John Everett Millais foregrounds the existential and phenomenological relation between the individual and the volume to which it has given birth, the Black Balloon campaign shows how this relation is no longer visible while remaining even more complex, more essential to our form of life. The Black Balloon campaign is obviously quite different both in terms of intent and aesthetics from the scene described by Sloterdijk. Rather than foregrounding the existential and phenomenological relation between the individual and the gaseous volumes they produce, the Black Balloon campaign instead foregrounds the relations between the issuing of volumes of gas into the atmosphere and the hidden networks of devices, practices, and technologies upon which this process depends. It gives shape to the connective materiality between the technical assemblages and atmospheres of modern forms of life. As an intervention designed to make explicit the composition of the atmosphere as an issue around which a particular affective-material public may emerge, the Black Balloon campaign fore-grounds the role and participation of atmospheric things in these publics, things which, while not quite alive, have an agentic quality issuing forth from their activity.[25]

BIG AIR PACKAGES

Such works make explicit the volumes of what Sloterdijk calls the "pneu-matic pact of modernity." By this he means that modern forms of life in-volve living with, storing, circulating, and disposing volumes of gas. This pneumatic pact is ever more discrete, hidden, and diffuse, in part because some of the more visible structures around which it was organized are dis-appearing, at least from contemporary urban landscapes in industrialised economies.[26] Of these, the gasometer is perhaps the most iconic. While appearing static, gasometers are dynamic structures, composed of an exo-skeleton within which a gasholder made of different sections can rise and fall as gas is pumped in or released. While the volume of the gasometer varies, the pressure of the gas in the domestic network can be maintained at a reasonably constant level by the weight of the dome that capped the structure. For much of their existence gasometers have been considered ugly structures despoiling the views and vistas of urban landscapes. Writing in *The Art World* in 1916, George Martin Huss remarked upon "the gas-tank nuisance," bemoaning the "ugly" and "repellent . . . monstrosities storing the supplies of illuminating gas" and asking "why not make these agreeable to look at, if we can do it by wedding Art and Utility?"[27] The *Art World* editors agreed, noting with particular displeasure a gas tank located near 130th street and Riverside Drive in New York. They asked instead that these "monstrosities" be sunk under ground, or designed "by some good architect along the lines of a Dome."[28]

Attitudes to gasometers have changed to some degree, not least because they are disappearing.[29] Many have been dismantled, but some have been put to new uses.[30] The gasometer in Oberhausen, just north of Düsseldorf, is a good example. Completed in 1929, it is 117.5 meters high, with a stor-age volume of 347,000 cubic meters. Used originally to store waste gases produced by blast furnaces in the surrounding industrial region, in addi-tion to higher-energy coal and coke gases, it worked by means of a pres-sure disc—upon which concrete weights were placed—that could move up and down the oil-lubricated and sealed walls according to the volume of gas in the holder. Having survived World War II largely unscathed, the gasometer was destroyed by fire before being fully rebuilt. Decommissioned in 1988, it was converted during the 1990s into an exhibition hall, becoming an important cultural center in the region.[31] From mid-March to late December 2013 the largest indoor sculpture in the world was installed in the Oberhausen Gasometer. *Big Air Package*, by the artists Christo and Jeanne-Claude, was proclaimed "the largest ever inflated envelope without a skeleton."[32] Sponsored by German industrial gas company Messer, and made from semitransparent polyester fabric supported by 4,500 meters of rope, *Big Air Package* was 90 meters high and 50 meters in diameter, with a volume of 177,000 cubic meters.[33] The envelope was kept inflated by two fans that maintained the pressure constant, with two airlocks allowing visitors to pass in and out of the sculpture.[34]

In *Big Air Package*, the technical problem and immersive experience of volume is aligned closely with the monumentality of architectural and aesthetic matters of concern. From between the outside of the envelope and the dark, oily inner walls of the gasometer, it is possible to sense, albeit only partially, the monumental: in the narrow confines of this gap it is also impossible to view the scale of the work in its entire height or diameter. Instead, the envelope appears like a fat, blunt missile housed in a silo. To enter the inflated envelope is to be overwhelmed by its proportions, to struggle to grasp a sense of the light, white emptiness extending around and upward from the body of the visitor until it becomes a translucent dome whose distance from the observer is never quite clear; unsurprisingly then, Christo compares the experience of being inside the envelope as akin to being in a cathedral (figure 5.3).[35] There are other comparisons to be drawn, of course. The experience of entering the work is akin perhaps to the experience of Andrée as he enters the inflated envelope of his balloon housed in the enclosure in Svalbard. It is akin to feeling a sense of double envelopment producing a spatiotemporal experience excessive of the volumetric.

The experience of immersion generated by works of such volume can be questioned. Christo and Jeanne-Claude's *Big Air Package* perhaps exemplifies what Peter Schneeman has identified as a trend toward the monumental in contemporary art, a trend linked to the "aesthetic language of 19th century memorials, the monumental canvasses of academic painting, [and] the grande machines," and redolent of nineteenth-century monumental exhibition spaces such as the Crystal Palace in London and the Grand Palais in Paris.[36] The extent to which this is a good thing is, of course, contested; as Henri Lefebvre reminds us, the monumental is a space

FIGURE 5.3 Inside *Big Air Package*, by Christo and Jeane-Claude, installed in the Oberhausen Gasometer, December 2013. Photo by author.

implicated in the production and reproduction of power on an architectural, gestural, and symbolic level. In the vein perhaps of Lefebvre's critique, noted earlier, a work such as *Big Air Package* reproduces a form of monumental volumetrics, similar to that of the cathedral, in which the subject is positioned as beholder of the magnificent enormity of power rendered in structural form. Schneeman takes a similar line when thinking about how, by generating feelings of being overwhelmed and enveloped, such works operate via "rhetorics of power" in which "the individual subject has to define a position against something that is bigger than his or her own mental and physical capacity, . . . becoming completely immersed, and thus losing consciousness of the self that needs distance to be constructed."[37] This is amplified by an atmospheric aesthetics of the sublime in both a scalar and a metaphysical sense: in the volumetric and voluminous dimensions of these works, the self submits itself to the experience of a "non-specific impression of a fundamental topic, a higher value, a reference system that hovers between the spiritual and the metaphysical."[38]

For Schneeman the aesthetic of a new monumentalism is characterized by "complete openness to symbolic associations and spiritual experiences" and to working with "undefined voids, using visual effects of the atmospheric."[39] Read through these terms, arguably *Big Air Package* celebrates as an end in and of itself the condition of both immersion in and envelopment by a vague atmospherics. In some ways, the critique that can be leveled at this work is similar to that made against the Pepsi Pavilion. This critique revolves around what Hal Foster calls the "ideological effects" of celebrating atmospheric immersion as an end in and of itself.[40] Of course, this critique is premised on the basis of a model of the subject as a clearly defined locus of agentic self-control in contrast to a vague, drifting, hazy atmospherics. It precludes the possibility that subjectivity itself may also be cloudy, atmospheric, and meteorological in its affects and percepts. If we take this seriously, however, it may be that such works as *Big Air Package* provide opportunities to appreciate the atmospheric as part of the condition of being immersed in an elemental milieu whose affects remain excessive of entities.

There is another critique that can be leveled at *Big Air Package*, however. In *Biogea*, Michel Serres writes, "To amaze the crowd and get himself talked about, an artist wrapped bridges, buildings, statues in public squares."[41] For Serres, Christo's work exemplifies a more general, and more problematic, obsession with the envelopment, or "wrapping," of the world in a layer of artifice and representation. By "wrapping" Serres means not only containment by physical fabric but also the ontological and epistemological imperatives to envelop the unformed, to represent that which can be sensed nonrepresentationally. He laments the emergence of a world "wrapped with words, sentences, images."[42] Serres dreams, instead, of "things freed of these packages, the way they presented themselves before finding themselves named."[43] Serres's dream is an alluring one: it is an invitation to think elementally like the wind, to move and feel and respond to forces and agencies that have not been enveloped by representational forms.[44] But it is based on a kind of naturalism, in which techniques of envelopment are defined automatically as somehow constraining. And yet, across so many domains of life it is the very process of wrapping and enveloping that generates a relation of difference through a process of partial enclosure. Envelopment can be more generative and enabling than

FIGURE 5.4 The view upward in *Big Air Package*, Oberhausen Gasometer, December 2013. Photo by author.

Serres suggests: especially if grasped in terms that are more akin to what he has described elsewhere as a process of generating volumes—such as boxes—through which the turbulent affects of the world can be sensed or made to resonate.[45] The point is to keep atmospheric envelopes open such that they do not become forms of volumetric power but remain volumes that have a spacious intensity that operates across and between bodies.

THINGS BECOMING VOLUMINOUS

Instead of rehearsing a blanket critique of wrapping or enveloping, it might be possible instead to experiment selectively with these processes in ways that foreground the very elemental forces affirmed by Serres. Here Sloterdijk's interest in spheres of life is useful: it can be read in a way that tempers Serres's disdain for envelopment. Considered more generously, envelopment is a process generating different forms and experiences of atmospheric volume in ways that are tensed between the volumetric and the voluminous. Experiments with the experience of envelopes of atmospheric volume can also be less monumental than Christo's efforts, however; it is not necessary to feel overwhelmed or overawed in order to experiment with this experience. Atmospheric volumes might then emerge as a sense of spacious intensity through different forms of relation and association involving bodies of varying sizes, speeds, and durations.

Work No. 200, Half the air in a given space, by the British artist Martin Creed, first shown in 1998, points to such possibilities. The instructions for the work are simple: "Choose a space. Calculate the volume of the space. Using air, blow up sixteen-inch balloons until they occupy half the volume of the space."[46] Creed's use of the balloon exemplifies his interest in transforming the use and meaning of everyday objects through encounters that render these objects strange. To some extent *Half the air in a given space* merely does what many of us have dreamed of doing, and some of us have done: to play around in a field of spheres. It affords opportunities for playing with the volume of discrete things and of the atmospheric affects that gather around these things. It is a volumetric exercise; the simple calculation of the dimensions of a space and the modification of that space on this basis. It is also a voluminous work in that it generates a relational environment in which visual depth and tactile distance are transformed through a form of spacious intensity. To be enveloped within the work is to have no view from above or outside. Bodies press in but can also be pushed away. The affects of this spacetime are, variously, hilarity, joy, and excitement, but also, in some cases, claustrophobia and anxiety. In Creed's work, then, the calculability of volume becomes a conditional constraint for generating an incalculable but palpable atmospheric field.

This sense of volume becoming voluminous is amplified in other works. *Scattered Crowd* is a work by the choreographer William Forsythe. First shown in 2002, it was installed in the theater at Hellerau, near Dresden, in October 2013. *Scattered Crowd* is composed of two kinds of balloon: helium-white and air-translucent. Each kind of balloon is tethered to its opposite by a thin line so that the space is composed of hundreds of oppositional pairs of random length. At Hellerau, a light mesh net formed a barely visible ceiling above which the helium balloons could not float. From the edge, then, there were just enough balloons to block any view into the distance, but not enough to make the space too clouded or too crowded. The work is visual, tactile, but also sonorous—heard before seen. Its intensive spaciousness is first sensed as sound, in the slow wave of a refrain becoming more voluminous as visitors move toward the doors.

Scattered Crowd is a work of choreography, and the balloons are what Forsythe calls "choreographic objects." The philosopher Erin Manning has written perceptively about how *Scattered Crowd* works to provide an "enabling constraint" for an ecology of atmospheric experience.[47] As she notes, the balloons themselves are not the point of the project but are volumes that allow for the creation of voluminous spacetimes of experimentation. The aim is to create volumes that evolve across an atmosphere whose variations are both generated and felt by multiple participants as they move with each other.

This is not about the dissolution of forms; walls and ceiling continue to mark out the limits of this space. Equally, the volumes of the space do not evolve naturally. The voluminous spaciousness of *Scattered Crowd* needs continuous tending and attention by those who install the piece. Helium escapes. Balloons descend and are replaced. Balloons also move, and cluster, both with the air currents in the space, and because people play with them. Tethers become entangled. Empty zones appear in the work, and redistribution is required. As a work of volume, therefore, *Scattered Crowd* is an ongoing production, a relational achievement. Within this work are zones of different density, regions in which things are a little more diffuse. At the same time, sometimes the movement of one balloon stands out as it sways and drifts on invisible air currents. The overall effect is a work whose spaciousness is always more than metric but registers through variations in the felt intensity of relations. As Manning explains, "The changing of the affective tone of spacetime is not willed by individual participants. It happens in a relational becoming: the room moves the participants to alter the composition of event-time. Here we see precision of proposition meeting unpredictability of event. To achieve a singularity of experience, the enabling constraints immanent to the proposition have to be both concise and open-ended. When it works, the whole atmosphere is moved."[48]

There are some obvious parallels here between *Scattered Crowd* and other works, among them Warhol's *Silver Clouds* (see chapter 3). Like *Scattered Crowd*, *Silver Clouds* is a choreographic experiment in which bodies and things commingle in producing an atmospheric volume. As in *Scattered Crowd*, visitors to *Silver Clouds* can touch, move through, and play with the balloons. However, unlike Forsythe's work, the balloons are not tethered to one another, nor can they be tended in the same way. As instructive, however, is the comparison between *Scattered Crowd* and Christo's *Big Air Package*. Both works experiment with volume in the twin senses of the volumetric and the voluminous, and in ways that unsettle depth of perception through a visual field lacking a horizon. Both are generative of atmospheric volumes in ways that echo some of the experiences of balloonists aloft. In rather different ways, each has the potential to produce the vague oceanic experiences described by balloonists such as James Glaisher and the sense of lightness and weightlessness invoked by Camille Flammarion and Albert Santos-Dumont (see chapter 2).

Such works are voluminous in another important sense, however: acoustic. To enter *Big Air Package*, housed as it is in the gasometer, is to encounter a zone of muted reverberations and momentarily strident echoes. It is to hear a continuous murmur of barely differentiated noise variously registering as voice, metal, footstep. It is to hear the periodic activation of the two air pumps as they keep the pressure relatively constant. To listen to *Big Air Package* is to hear the sound of the work of gaseous modernity, to hear its reverberation as an echo. To enter *Scattered Crowd* is also to become enveloped within a sonorous atmosphere, the volume of which waxes and wanes as part of the loop of a refrain punctuated by a pause of sorts.[49] To listen with and within *Scattered Crowd*, is, following Serres, to sense the refrain of an atmospheric volume as its intensive spaciousness shifts and alters through the compositional relations between moving bodies. *Scattered Crowd*, in other words, allows us to hear variations in atmospheric volume as much as to see or feel them, to hear the sounding of volume as a swirl of relations and movement. *Scattered Crowd* reminds us that the voluminous experience of atmospheric envelopes is not only visual or kinesthetic; it is characterized as much by degrees of sonic intensity as by spatial extension.

There are important differences, however, between *Scattered Crowd* and *Big Air Package*. While *Big Air Package* is enveloping, it positions the visitor as a kind of spectator. The invitation is to experience the voluminous qualities of lightness and perhaps weightlessness: to lie back, and feel as if you are floating in a vague white sphere of light. But there is little sense in which the presence of the visitor makes a tangible difference to variations in the volume of the envelope (figure 5.5). *Scattered Crowd*, and in important ways both *Silver Clouds* and *Half the air in a given space*, in contrast, are very different. They foreground the participation of different bodies—human and nonhuman—in the relational production of atmospheric volumes. Critically, unlike Christo's *Big Air Package*, Forsythe's *Scattered Crowd* manages to produce volumes of significant size without

FIGURE 5.5 William Forsythe, *Scattered Crowd*, 2002, installed at the Festspielhaus, Hellerau, October 2013. Photo by the author.

necessarily rehearsing the spectacle and sensation of monumental power associated with the architect as visionary.

Christo's work is an envelope that generates the sense of being enveloped in a diffuse light, white spaciousness. Composed through the shape and capacities of discrete envelopes, *Scattered Crowd* produces an envelope of atmospheric spaciousness composed of relations and associations. In the case of both works the atmospheric thing is not the individual balloon; rather, it is the palpable sense of an atmospheric volume forming as a cluster of associations that never take shape as an object. Instead of positioning us as rapt observers of the sheer all-enveloping breadth and depth of the volumes we inhabit, *Scattered Crowd* points instead to our modest implication in the emergence of the circulations and associations that shape the voluminous spacetimes within which we move. *Scattered Crowd* invites us to think of spacetimes in ways that lie somewhere between Sloterdijk's associational forms and Serres's unwrapped turbulent swirls and eddies. In this work, envelopment is the process that holds us between the form of an entity and the unformed sense that something atmospheric is taking place.

Like the volumes encountered in chapter 4, *Scattered Crowd* may well be a volume whose allure has much to do with the way in which "every surface communicates."[50] Indeed, the origin of the term "volumes" lies in the idea of a scroll of paper or material that can be unwound as a surface, with the meaning of size and extent deriving from the idea of the sheer mass of such paper, particularly in the form of a book. However, just as the affects of such books are not limited to what lies on the surface of a page, the allure of atmospheric volumes is not limited to something that emanates from surfaces. Instead, we might think of how the sense of volume as spacious intensity emerges through a process that Sasha Engelmann calls "surfacing."[51] In this respect, the surfaces of the volumetric entities of which *Scattered Crowd* is composed contribute to the production of a voluminous atmosphere that becomes palpable, or surfaces, as a cloud of elemental variations.

DIFFERENT VOLUMES

The volumes of atmospheres are not just volumetric: they have a more-than-geometric quality grasped as a differing sense of the voluminous produced across and between bodies and their moving relations. Differentiating volume is necessary in order to avoid thinking of it as a dimension of atmospheres always already there, a dimension existing prior to the use or occupation of space. Volume, then, is not merely the specious empty space of which Henri Lefebvre is so critical, nor merely the product of commodification and abstraction. Atmospheric volumes are emergent or evolving properties of the relations between bodies, human and nonhuman, and their capacity to change. They hold open the envelope as a form, and envelopment as a condition, in ways that are not circumscribed by assumptions about particular spaces. Where the volumetric specifies the shape and properties of particular spaces, the voluminous remains vague as a spacious intensity in which different bodies can precipitate or move. The importance of thinking about such works is, following the insights of Luce Irigaray, that it might facilitate experiments with ways of thinking about an envelope that cannot be "circumscribed because it's open."[52]

This is not to dismiss the importance of envelopment as a technical process for engineering experiments with a differentiated sense of volume. Nor is it to dismiss the volumetric as some kind of masculinist register of governmental spatiality. In many ways it is vital that envelopment renders volume explicit in geometrical terms through scientific experiment and, equally, through the engineering and construction of architectural spaces. As the Black Balloon ad campaign illustrates, this can have ethico-political value insofar as it gives a shape of sorts to atmospheric processes that have become increasingly hidden from view in the worlds that many of us inhabit. Yet, at the same time, the volumes of the atmospheric can be sensed in and through fields of experience via different techniques and practices of envelopment as ethico-aesthetic experiments. The important point is that experiments with envelopment help us understand volume not as an a priori property or quality of empty metric space but as a relational and fragile achievement whose intensive spaciousness is felt in atmospheres of different kinds.

6   SOUNDING

To think of the volume of atmospheres is to invoke a sense of the sonorous. In many ways, the question of how atmospheres can be sounded is critical to the problem of how atmospheres are sensed, even when this sound is of a frequency or wavelength beyond the limits of human hearing, and even when whatever is sensing this sound is not human.[1] It is hardly surprising, perhaps, that early aeronauts remarked frequently upon the sonorous qualities of the atmosphere in which they traveled, paying particular attention to how sounds were modified and modulated by the meteorological characteristics of the medium through which they moved. James Glaisher, notably, wrote that "fog is much more sonorous than dry air, and collects sound with such intensity, that whenever, in passing through a cloud, we have heard a band playing in a town beneath us, the music always seemed to be close at hand."[2] And, of a flight over London, he wrote, "When one mile high the deep sound of London, like the roar of the sea, was heard distinctly; its murmuring noise was heard at great elevations,"[3] while at four miles, "the roar of the town heard at this elevation was a deep, rich, continuous sound—the voice of labour."[4]

The qualities of the echo aloft also fascinated aeronauts. Joseph Louis Gay-Lussac was a French physicist and chemist whose interest in balloon flight paralleled his scientific investigation of the properties of gases in the laboratory and in the air.[5] According to some accounts, on one flight Gay-Lussac brought a "speaking-trumpet" with which to experiment with sound. During the ascent, so one version of its progress goes, Gay-Lussac noted that the "voice, through a speaking-trumpet, was re-echoed most perfectly from the earth, even at the greatest elevation; and the time of the return of the echo so well coincided with their height, increasing in quickness as the latter diminished."[6] Apparently Gay-Lussac even suggested this interval could be used as an accurate measure of altitude, while also noticing that whenever he spoke into the trumpet "a slight undulation of the balloon was perceptible."[7] Or take the balloonist, astronomer, and writer Camille Flammarion. Like Gay-Lussac, Flammarion considered the echo a measure of altitude. But he noted also its aesthetic qualities, its "beauty" above a "wide sheet of water."[8] Of a flight near Paris in 1867, Flammarion wrote, "I shout: the sound returns as an echo, after a lapse of six seconds. It would be interesting to ascertain whether the vertical velocity of sound is equal to its horizontal velocity in the air, and if the echo is really returned from the plane beneath. . . . I was much struck by the vague depth of the echo: it appears to rise from the horizon, and has a curious tone, as if it came from another world."[9]

Like others, Flammarion was also fascinated by the strange and profound experience of echoless silence often characteristic of being-in-the-air. Of the experience of being at eleven thousand feet, Flammarion wrote, "Absolute *silence* reigns supreme in all its sad majesty. Our voices have no echo. We are surrounded by a vast desert. The silence which reigns in these high regions of the air is so oppressive that we cannot help asking if we are still alive."[10] The absence of an echo seemed to reaffirm, for Flammarion, the existential reassurance provided by the return, resonance, or reverberation of sound. Without an echo, there was no palpable limit to the space in which the self was immersed, nor any way of taking the measure either of that space or of the self. Frances Dyson argues that when "immersed in sound, the subject loses its self, and in many ways, loses its sense."[11] But the converse is also true: without sound, the spacing of the self can seem too expansive. The absence of any echo points to the exposure of life to something beyond the reassuring envelope of the atmosphere, an alluring if also subliminal realm in which the propagation of sound, as both a physical and existential process, is impossible. For Flammarion, the "imposing" silence pointed to whatever lay beyond the atmosphere, to the absence of a sonorous, audible signal, providing a "prelude of that which reigns in the interplanetary space in the midst of which worlds revolve."[12]

For the balloonists above, then, the echo was both an experiential phenomenon and a technique for experimenting with and within the atmosphere: in that sense it held together the volumetric and the voluminous dimensions of atmosphere. Its duration provided an index of measurable extent and calculable depth, while also generating a sonorous affective-aesthetic experience. Through minor experiments with the echo and its absence, these balloonists were *sounding* atmospheres. To sound is to render something sonorous: to give it expressive, audible force, making it available for sensing. We often tend to think of this as a process involving distinctively animal capacities. To sound, in these terms, is to vocalize something (but not necessarily to form it in words) that can be heard in the sensory loops and canals of animal bodies.[13] But sounding does not necessarily involve the animal body. Sounding, as a process of making audible, is abroad in the world. As Michel Serres reminds us, "In myriads, things cry out. Often deaf to alien emissions, hearing is astonished by that which cries out without a name in no language."[14] Furthermore, the source of this cry may not be a body or an entity but something far more diffuse: the buzzing, background noise of the world, or, beyond this, the refrain of the between of things, of a kind of original sonorous differentiation of the universe. Sounding begins, we might say, on the "threshold of the echo."[15]

To sound is also, however, to test: it is to explore the properties, qualities, or extent of a body of liquid. Deriving not from the sonorous but from the old English word *sund*, meaning sea, sounding in a maritime or navigational sense involves a "bathymetric" process of assaying the depth of the water in which a vessel floats. In its most simple iteration, it involves dropping a line over the side of a vessel in order to map the floor of a body of water. Today it is more likely to involve the use of an underwater echo (known as sonar). Derived from the maritime, this understanding of sounding is also used to describe the process by which the dynamics of the meteorological atmosphere are disclosed. Sounding here involves the use of the balloon as a device with which to ascertain, remotely, and in situ, variations in atmospheric processes. The difference here, of course, is that the depth, or height, of the atmosphere (a least as taken from the ground) is not mappable in the same way as the sea. Nor is there a solid surface at the top of atmosphere against which to bounce a signal. Equally, with some exceptions, atmospheric sounding does not involve the use of tethered lines. Instead, for the most part it involves the release of free meteorological balloons into the atmosphere.

Taken together, these twin senses of sounding—as the process of making audible and making explicit—disclose atmosphere as a medium for particular kinds of media experiments. These are experiments linking the affective atmospherics of communication media with the idea of the atmosphere as an elemental milieu whose variations are messages that can be discerned. They are experiments that stretch the envelope of possibility for transmitting and receiving messages as part of the engineering of increasingly global assemblages of communication. They are media experiments in which the dynamics of the meteorological atmosphere are disclosed through assemblages of objects and devices in the air and on the ground. But they are also media experiments that, in certain circumstances, remind us that atmosphere is not just that which can be sensed through sounding but is the very condition that makes particular kinds of sensing possible.

SOUNDINGS I: ECHO

In the late 1950s, NASA, AT&T, and Bell Labs began collaborating on the use of passive satellites for radio transmission under the name of Project Echo. The satellites used in the experiment were large, reflective balloons launched into low orbit to act, in effect, as huge spherical mirrors against which signals transmitted from one site could be bounced toward another. Something of the scope of the project, and of the structure of feeling within which it was undertaken, can be gleaned from *The Big Bounce*, an educational short film made about the experiment.[16] The soundtrack of the film captures the atmospheric "tone of the times" within which the project was undertaken.[17] The eerie music and the slightly portentous narration are strangely suggestive of something otherworldly, of the unfolding of an experiment mixing science fact and science fiction. The narration begins, "The place, a hilltop in New Jersey. The setting, giant antennas, like monstrous eyes and ears, straining, watching, waiting." After outlining the nature of the project, the narrator observes that if the experiment works, "it will be the first time voice has traveled from the earth up to a man made moon, and back to earth again." The narrator then pauses to pose a question—"Who cares about bouncing messages off a space balloon?"—before outlining the need for "more and different systems" through which to broadcast and continuously transmit radio and television signals.

Project Echo had its origins in various speculative proposals during the immediate post–World War II period about the possibility of using balloon satellites as technical, and to some extent geopolitical, devices.[18] In 1946, the RAND Corporation published its very first report, titled *Preliminary Design of an Experimental World-Circling Spaceship*, in which the possibility of using a balloon was considered briefly in the context of a range of other proposals.[19] In 1947 a follow-up report examined these issues further, proposing the use of satellites as reconnaissance devices. In the report, one of the members of the RAND team, James Lipp, had outlined a system of orbiting satellites and ground-based relay stations via which data could be transmitted beyond line of sight distances.[20] This proposal echoed, without acknowledging, ideas Arthur C. Clarke had articulated in a 1945 article about the value of a network of orbiting satellites for the global transmission of radio signals.[21]

In the early 1950s an electrical engineer at Bell Labs, John Pierce, also began thinking about a system of communication satellites orbiting the earth. Considering these ideas too far-fetched for professional scrutiny, he chose to publish them initially in popular science fiction magazines. However, by the mid-1950s, he began publishing openly about the design of different kinds of satellites, ideas that would shape his practical experiments at Bell Labs. These experiments drew upon and drew together an assemblage of emerging technologies, including the transistor, the traveling wave tube, solar cells, and early kinds of laser, in addition to a type of antenna called the horn antenna, which allowed microwave signals to be focused and received.[22] Combined with experiments with rocket engines then being undertaken by the Jet Propulsion Lab at NASA, the work at Bell Labs pointed to the development of satellites that could facilitate the transmission of strong and focused signals.[23] The ultimate goal of these experiments was the design and launch of active satellites that could both receive and transmit signals.

Project Echo was intended as a kind of incremental step toward the development of active satellites. It remained dependent upon a passive device, however, which could only receive a signal from a ground source and reflect it back to a large ground-based antenna. Within this constraint, the aims of Project Echo were fourfold:

1 To demonstrate two-way voice communication between the east and west coasts by microwave reflection from the satellite.
2 To study the propagation properties of the media, including the effects of the atmosphere, the ionosphere, and the balloon.
3 To determine the usefulness of various kinds of satellite tracking procedures.
4 To determine the usefulness of a passive communications satellite of the Echo I type.[24]

As figure 6.1 indicates, the project involved a distributed assemblage of devices, including a transmitter in California and a receiver at Holmdel, New Jersey. At the center of this assemblage was, however, a balloon. This had been designed at Langley Air Force Base by William O'Sullivan, who conceived it initially as a sounding device of sorts, with which to "measure the density of the air in the upper atmosphere and thereby provide aerodynamic information helpful in the design of future aircraft, missiles, and spacecraft."[25] But O'Sullivan was aware also of the contribution such balloons might make to the development of more extensive communication systems through which the technological, ideological, and affective dimensions of Cold War geopolitics might be amplified. Appearing before the House Select Committee on Science and Astronautics in April 1958, he enthused about the feasibility of launching such a balloon into orbit, claiming it "would reflect radio signals around the curvature of the earth using frequencies not otherwise usable for long range transmission, thus mostly increasing the range of frequencies for worldwide radio communications and, eventually, for television, thus creating vast new fields into which the communications and electronics industries could expand to the economic and sociological benefit of mankind."[26] To reinforce his point, during his visit to the committee he inflated a twelve-foot foil-covered balloon satellite in the Capitol Building.[27]

O'Sullivan faced particular challenges in designing and fabricating the envelope for this balloon: he needed a material light and compact enough to be launched into orbit in a rocket, strong and flexible enough to be inflated in orbit, and durable enough to withstand the rigors of heating and cooling in space. It also had to be reflective enough to be tracked from the ground. His solution was a laminate of two materials. The first was a new kind of very strong plastic film, Mylar, which had been developed recently by the DuPont Corporation and was also being used as audio recording tape and in temperature-resistant food storage bags. The second material was a thin aluminum foil developed by the Reynolds Corporation and ap-

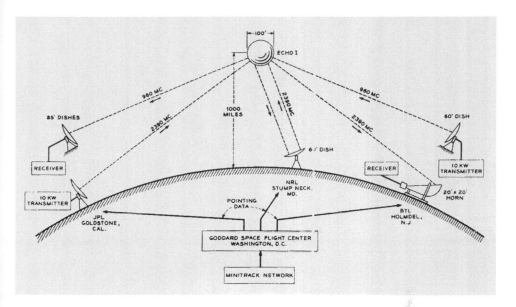

FIGURE 6.1 General features of the Project Echo Experiment. From NASA Technical Note D-1127, "Participation of Bell Telephone Laboratories in Project Echo and Experimental Results," authored by William C. Jakes, Bell Labs.

plied through a vaporization technique. Even before the balloons were launched into orbit they presented alluring reflective surfaces, and they conjured some of the qualities of the envelopes with which those involved in later experiments at EAT would work. Inflated for testing in a hanger in North Carolina, the Echo balloons seemed to anticipate the reflective aesthetics of envelopment that would later take shape in the mirror dome of the Pepsi Pavilion, and, in a different way, in the foil-covered envelopes of Warhol's *Silver Clouds*. These aesthetics can also be situated in relation to the wider fascination in the United States, during the late 1950s and early 1960s, with reflective metallic materials. This was a period during which aerial devices of all kinds, particularly military airplanes (to say nothing of chrome-adorned cars), were fabricated from aluminum polished to the gleaming shine of what Mimi Sheller calls "light modernity."[28] In the period before disguise and stealth became critical to aerial power, such new reflective materials provided surfaces for projecting an affective and geopolitical aesthetics of confidence.

O'Sullivan's balloons became the model for the satellites used in Project Echo. These were larger than those with which O'Sullivan had experimented

FIGURE 6.2 Static inflation test of Echo satellite in Weeksville, North Carolina, June 28, 1961. Photo: NASA.

initially, and, as such, presented new technical challenges (see figure 6.2). During testing, the seams of the balloon were revealed to be too weak. In addition, a series of test launches into the upper atmosphere were undertaken to examine how a second, strengthened balloon would cope with the rapid inflation of the envelope at this altitude. The first of these launches resulted in the destruction of the balloon when residual air in the folded envelope expanded explosively in the upper atmosphere. The problem was solved by using an inflation agent that transformed slowly but directly from a solid to gas (sublimation) and by making three hundred holes in the envelope to allow air to escape.

One of the key difficulties faced by the engineers was rather more mundane, and involved the problem of the relation between folding, envelopment, and inflation. Specifically, the problem was how to fold the balloon in such a way that it would fit inside a specially designed canister released from a rocket while also being free to inflate without tearing. Apparently, the solution to the problem was inspired by the design of a plastic rain hat owned by the wife of one of the project team members. The work of learning to fold the fabric was described thus:

> At Langley, Kilgore gave the hat to Austin McHatton, a talented technician in the East Model Shop, who had full-size models of its fold patterns constructed. Kilgore remembers that a "remarkable improvement in folding resulted." The Project Echo Task Group got workmen to construct a makeshift "clean" room from two-by-four wood frames covered with plastic sheeting. In this room, which was 150 feet long and located in the large airplane hangar in the West Area, a small group of Langley technicians practiced folding the balloons for hundreds of hours until they discovered just the right sequence of steps by which to neatly fold and pack the balloon.[29]

Folding and refolding were critical parts of the process of fabricating such envelopes. Working with the capacities and limitations of this lustrous new material, and of tending carefully to the way in which it behaved under a range of conditions, were crucial processes through which new and alluring technologies of atmospheric experiment emerged.

Despite its first official launch ending in disaster, as a technical experiment Project Echo was a success insofar as it demonstrated the technical feasibility of passive satellite communication, becoming the first object in space against which a signal from Earth was transmitted and bounced.[30] After a second trouble-free launch in August 1960, the Echo 1 balloon entered orbit around the earth at a speed of about sixteen thousand miles an hour and at a height of approximately one thousand miles. During its first orbit the balloon relayed a message from President Eisenhower from California to New Jersey: "It is a great personal satisfaction to participate in this first experiment in communications involving the satellite balloon known as Echo. This is one more significant step in the United States' program of space research and exploration. The program is being carried forward vigorously by the United States for peaceful purposes. The satellite balloon, which has reflected these words, may be used freely by any nation for similar experiments in its own interest."[31]

The Echo balloon continued to orbit for well over eight years, during which time it facilitated experiments with satellite-borne two-way conversations and fax transmissions. And while the launch of the Telstar satellite not long afterward rendered the Echo satellite technically obsolete, it did have some important results, with both NASA and AT&T claiming that it was a critical experimental step in the development of global communication systems. In addition, as it continued to orbit for eight years, "the satel-loon allowed scientists to measure accurately, for the first time, the density of the air in the far upper atmosphere," and to ascertain how this density was modified by solar activity.[32] Project Echo was also followed up by a series of similar launches under the name of Explorer, launches that generated further data about atmospheric density. In addition, in 1966, the Pageos balloon was launched as part of a five-year project involving twelve mobile tracking stations in order to provide precise data about the shape and size of the earth.[33]

The Echo balloons were orbiting satellites rather than more conventional atmospheric sounding balloons. But they can be understood as sounding experiments of a sort insofar as they were critical to efforts to gauge, and in some sense stretch, the limits of the envelope of modern forms of communication. The echoing, or bouncing, of the signal against the reflective surfaces of the Echo balloons was a critical moment in the emergence of contemporary infrastructures of atmospheric media.[34] Even if these infrastructures were not all about the transmission of the voice, in the bouncing of Eisenhower's message against the envelopes of the Echo balloons, the dream of a global media atmospherics was sounded with particular clarity.

Project Echo was a pilot experiment for the development of a distinctive kind of media infrastructure that was simultaneously technical, geopolitical, affective, and atmospheric. The launch of the Echo balloons pointed to the emergence of assemblages of media transmission that could connect distant places in novel ways but also, in doing so, could modulate the affective atmospherics of the Cold War.[35] Against the backdrop of the launch of the Soviet Sputnik satellite a few years earlier, the Echo satellites became objects of affective allure in a world becoming sensitized to the presence of new kinds of devices orbiting the earth. The psychological and political benefits of such a satellite program had been identified in the 1945 RAND report. Anticipating the surprise that would accompany the Sputnik program a few years later, James Lipp wrote in the conclusion to the RAND report, "One can imagine the consternation and admiration that would be felt here if the United States were to discover suddenly that some other nation had already put up a successful satellite."[36] The symbolic and affective value of launching something visible into near space had also been earlier anticipated by a chemist, Aristid von Grosse of Temple University, who, in a report for the Truman administration, had "recommended orbiting an inflatable balloon that would, to the naked eye, appear as an 'American Star' rising in the West."[37] The designer of the Echo balloon, O'Sullivan, had been similarly enthusiastic about the political significance of his satellites. He had earlier proposed launching into space a twelve-foot inflatable sphere known as Beacon for the sole purpose of getting some kind of object into orbit that might be visible to the naked eye over the USSR.[38]

It is hardly surprising that even the testing phase of Project Echo generated interest among a public primed to anticipate the possibility of seeing new kinds of artificial things in orbit. When the balloon had exploded during the process of inflation on an earlier test launch, thousands of pieces of the aluminized Mylar floated back into the atmosphere, "reflecting the light of the setting sun" and creating "sensational flashing lights" in the sky all along the Atlantic seaboard.[39] Rather than reporting the failure, press releases about the launch emphasized the success of launching objects into space, and newspapers featured headlines such as "Earthlings Stirred by NASA Balloon, Awesome Sight in the Sky."[40] Such interest was heightened once the Echo balloon entered orbit. Visible from the ground, these orbits became the focus of curiosity and wonder, amplified by the balloon's appearance in newspapers and magazines.

SOUNDINGS 2: GHOST

Later in the decade, on October 1, 1968, the National Center for Atmospheric Research, based in Boulder, Colorado, issued a press release under the following heading: "GHOST Balloon Completes One-Year Flight." The press release announced that a "ten foot plastic balloon, launched from New Zealand in September 1967, has just broken all previous balloon flight-duration records by staying in the air for one year."[41] During that year the balloon, fabricated from Mylar, had circumnavigated the southern hemisphere twenty-five times at a constant height of about 52,000 feet, its signals tracked by various stations.

The record-breaking flight was made by one of over eighty balloons launched from Christchurch, New Zealand, as part of the GHOST (Global Horizontal Sounding Technique) project (see figure 6.3). Involving a scientific collaboration between NCAR (along with other agencies in the United States) and agencies in New Zealand, the central aim of the GHOST project was to test the feasibility of using balloons to provide accurate real-time information about the atmosphere in ways that were previously impossible. In effect, the balloons were intended to act as "roving weather stations

FIGURE 6.3 Superpressure balloon in Christchurch, New Zealand hangar, circa 1968. Part of the GHOST Program. Source: National Center for Atmospheric Research Archives, Vincent Lally Papers.

which can collect atmospheric data to be used in global weather forecasting." This data was to be used, the press release stated, to aid the development of "numerical models of the atmospheric general circulation, which will be used to forecast weather by making it 'happen' in the electronic circuits of a computer faster than it happens in the real atmosphere."[42] The team at New Zealand was led by Vincent Lally, who had for some years experimented with balloons as devices for measuring atmospheric phenomenon. Instrumental in setting up the balloon research division of NCAR, Lally had developed, among other things, superpressure balloons that could maintain a constant altitude by floating at a specified atmospheric density.

Atmospheric data had been collected via vertical balloon sounding for decades. Indeed, the meteorological atmosphere was gradually disclosed by the use of balloon sounding. The first experiments with unpiloted balloons as devices for studying the upper regions of the atmosphere were undertaken in France in 1892 by Gustave Hermite and Georges Besançon, who launched recoverable recording instruments, including a thermometer, a barometer, and a hygrometer. Very quickly, the use of such instruments facilitated the identification of the stratosphere, with a range of other balloon types devised in the years following. A persistent problem faced by those who used such balloons, however, was how to recover them once they returned to earth. The invention of the radiosonde in 1929 changed this. Involving the addition of a radio transmitter to weather balloons, the radiosonde allowed the transmission of real-time data about wind, pressure, temperature, and humidity.[43] Equally, the development of new plastics from which to construct envelopes facilitated the operation of more extensive scientific ballooning during the 1950s and 1960s. The GHOST project emerged in this context, as an ambitious experiment in horizontal sounding, via which data about wind, temperature, and humidity could be collected as balloons circulated in the upper atmosphere.[44] The superpressure balloons it deployed were fabricated from strong, inelastic plastic. They were also equipped with transmitters and, importantly, with "cut-down" devices that could be triggered should the balloon drift over geopolitically sensitive territory or through the wrong airspace.

To reduce the likelihood of such incursions, a project launch site was established in Christchurch, New Zealand. Lally, his team, and indeed some of their families, moved to Christchurch. As with Project Echo, much of the work of the GHOST project involved devising ways of fabricating, inflating, and launching the balloons. Equally, the families of the project scientists and technicians sometimes became involved in the GHOST operations, albeit in ways that were perhaps more participatory than with Echo (figure 6.4).

Like Project Echo, GHOST also involved the development of a distributed assemblage of devices for sounding and listening, but one that was more obviously international in its scope and level of cooperation. Once they were launched, the work of sounding the upper atmosphere involved listening out for the balloons, each of which was identified by the distinctive signal (of dots and dashes) it had been assigned. This was facilitated via a network of listening posts and tracking stations scattered across the southern hemisphere at sites in Angola, Antarctica, Argentina, Australia, Brazil, French Polynesia, Mauritius, South Africa, and Zambia.

On one level, this experiment in sounding and listening was a technical operation involving the tracking of technical objects. But the experience of working with balloons released independently into the atmosphere is also an affective one. Once released, these devices become more spectral than the title of the project suggested. The movement and trajectories of the balloons faded in and out of the range of different listening posts, variously withdrawing into and reemerging from an undisclosed atmospheric zone, particularly over Antarctica. Lally's notes from the period hint that the periodic disappearance and surfacing of the balloons was a process experienced affectively by scientists as they listened for signals of the entities they had released into the atmosphere. In some cases the signal was difficult to discern, with the result that the status of various balloons sometimes remained shadowy and approximate at best. On February 8, 1966, for instance, Lally noted that he had "found mysterious balloon at 15.022," and that this was "GHOST of M." On May 8, he noted that the team was "unable to hear FFF today except faint background—apparently deep in Antarctic." On May 14, his notebook read, "Long lost 23206 FFF came back!!!"[45] For Lally and his team, then, sounding not only involved using the balloon as a device for sensing in situ the atmospheric field in which it moved. Sounding was experienced as an affective process of distributed listening for something spectral: a form of remote sensing involving attuning to the signal of superpressure balloons as they called out their position while circumnavigating the southern hemisphere.

Lally imagined GHOST as the precursor to a more ambitious project, which would include "5000 to 10,000 Ghost balloons, roving freely above the globe at predetermined levels of the atmosphere." Because it would simply be too costly, too logistically difficult, and too politically complicated to

FIGURE 6.4 GHOST balloon launch in Christchurch, New Zealand, circa 1970. Source: National Center for Atmospheric Research Archives, Vincent Lally Papers. https://opensky.ucar.edu/islandora/object/archives%3A2031.

try to build a network of ground-based receiving stations, data from these balloons would not be transmitted directly to the ground but would be relayed by "two or more earth-orbiting satellites."[46] The overall aim would be to provide a system for generating immediate, real-time information about the state of the atmosphere, data that could be used by increasingly powerful computers to generate more reliable weather forecasts. Framed by the geopolitical tensions of the Cold War and the technological optimism of the postwar period, the GHOST project was to be a precursor to a distributed assemblage of devices, including balloons, satellites, and sea buoys, through which the dynamics of the meteorological atmosphere could be sounded on an ongoing, long-term basis.

Lally's hope was that the requirement to understand the dynamics of the weather would generate the imperative to form new kinds of international cooperation. To some extent these hopes were realistic: a range of cooperative experiments were undertaken during the 1960s and 1970s, generating a system of progressively greater global reach for understanding the dynamics of the atmosphere and of weather. Nevertheless, in notes for a lecture on the promise and problems of a project such as GHOST, NCAR's Vincent Lally wrote that the acquisition of upper air data "turns out to be a global problem and a sticky political problem. However, we know the problems and have solutions in mind. Let's hope that in the next few years we shall be negotiating for the placement of balloon launch sites instead of the removal of missile launch sites."[47] Indeed, even over friendlier territories the GHOST balloons became objects around which the geopolitical atmospherics of the Cold War could crystallize. According to the *Christchurch Star* of August 1969, the population of the small Queensland town of Millmerran mistook one of these balloons for a UFO. A Canberra bomber investigated the sighting, and confirmed that the object was "a huge transparent bubble."[48]

As an experiment in meteorological sounding, the GHOST project was an exercise in making the dynamics of the atmosphere explicit. But it is also a reminder that these projects are media experiments of a sort because they bring the dynamics of the atmosphere within the orbit of a range of sensing devices and experiences. As Jussi Parikka has observed, "Practices of meteorology are mediatic techniques that give a sense of the dynamics of the sky."[49] GHOST anticipated a network of listening devices for realizing the promise of making more effective prediction of the weather available in everyday broadcast media. However, where Echo balloons were premised on the possibility of rising above the atmosphere in order to expand the envelope of atmospheric media nearer to the ground, the GHOST balloons moved with the trajectories of the atmosphere, and particularly with those of the stratosphere. As such, they point to something else: to the possibility of using the movement and variations in the atmosphere as infrastructural trajectories for the distribution of devices, ideas, and affects.

Sounding travels across and between GHOST and Echo. In the GHOST project, sounding is a technique for assaying the meteorological dynamics and properties of our planetary atmosphere. In Project Echo sounding is the process through which voice transmission provides an indicator of the technical viability of new forms of infrastructure that facilitate the development of envelopes of atmospheric media in terrestrial lifeworlds. We might say, then, that sounding is the both the process by which the dynamics of the atmosphere are disclosed and the process by which the limit, or envelope, of atmospheric sensing is stretched.

But how far might this envelope be pushed? Camille Flammarion, noted at the outset, speculated about what exists beyond the realm of the air or atmosphere. He wondered about how the silence of the "vast desert" beyond might move us to ask other questions about the limits and thresholds of sounding atmospheres. Questions about what it means to sound something whose origin is beyond the atmospheric envelopes in which life on earth is immersed. As it turns out, the balloon has for quite some time provided a platform for undertaking experiments designed to address such questions, and in particular through research on cosmic rays. During 1911 and 1912 the Austrian scientist Victor Hess made a series of balloon ascents to explore how radiation ionized the earth's atmosphere. On one of these ascents he realized that even during a solar eclipse the process of ionization continued within the atmosphere, penetrating through its lower layers but remaining significantly greater in the upper atmosphere. Hess concluded from this that the source of this radiation must be extraterrestrial. This "discovery" was not universally accepted, however. Further balloon experiments during the 1920s and 1930s confirmed the existence of cosmic rays, including experiments by Robert Millikan, who used unmanned sounding balloons, having acquired experience of their use during World War I.[50] It was Millikan who gave the name "cosmic rays" to such radiation.

The balloon in this context provides a platform for sensing something whose origins are extraterrestrial. The properties of cosmic rays continued to be investigated in this way throughout the twentieth century, remaining central, for instance, to the US Skyhook and Strato-Lab programs in the decades after World War II.[51] Crucially, the aim of such experiments was not to sense the dynamics of the atmosphere; instead, the atmosphere becomes a medium for sensing something whose origin is excessive of atmosphere, something that is sensed only through its capacity to generate a form of elemental perturbation in a medium such as air or water. We can think of this as another form of sounding, a form of sounding the force of something excessive of atmosphere.

Another experiment allows this to be probed further. While it does not involve a balloon, the site at which it took place, and the device used, were central to the infrastructural assemblage that facilitated Project Echo. The device in question is the horn radio antenna at Holmdel, New Jersey, constructed in 1959 as part of Project Echo, and taking its name from a parabolic horn-shaped structure with a curved aperture at its open end. The antenna is about fifty feet long and can be rotated such that it points toward any part of the sky. Critical to its operation in Project Echo was the use of a MASER (Microwave Amplification by Stimulated Emission of Radiation) that allowed very weak microwave signals to be amplified without adding any noise.[52]

In the early 1960s, two physicists at Bell Laboratories—Arno Allan Penzias and Robert Woodrow Wilson—were tasked with the rather routine job of measuring the sensitivity of the Holmdel antenna, with the agreement that they could then use it for astronomical observation.[53] During their work, however, Penzias and Wilson encountered a problem. Try as they might, they could not eliminate a persistent and steady noise continuously present at the frequency of microwaves, something like the "hiss that an old FM receiver might have made with an unused channel."[54] Penzias and Wilson were perplexed, but continued trying to identify the source in order to eliminate it from their calibration of the telescope. Examining the antenna, they removed pigeon droppings and got rid of the pigeons (by shooting them) in the hope that either may have been the source of the signal, but with little effect. The hissing noise continued, apparently without any discrete source, "and seemed to be coming from somewhere outside the atmosphere."[55]

After eliminating the possibility of any human source for the signal, including nearby New York City, Penzias and Wilson eventually began to speculate that its origins were extraterrestrial. Consulting colleagues, they came to the conclusion that the noise was the signature of cosmic microwave background (CMB) radiation. The existence of CMB radiation had been theorized for some time as the thermal signature of the initial generative event from which the universe emerged (now commonly known as the Big Bang). The oldest light in the universe, sometimes known as relic radiation, CMB radiation is really only detectable in the microwave range by the kinds of sensitive antennas with which Penzias and Wilson were working. Another team was searching for such radiation at the time Penzias and Wilson made their discovery, and after consulting them, both teams published simultaneous notes.[56] However, notwithstanding the fact that theirs was an accidental discovery, Penzias and Wilson were subsequently awarded a Nobel Prize for their work.

It is easy to understand this as a physics experiment even if the details of the physics are difficult to grasp. But no less than Project Echo and the GHOST project, this experiment, perhaps like all astronomical experiments, can also be understood as a media experiment—an experiment that poses the question of how forces excessive of terrestrial envelopes of atmospheric experience can be disclosed through making a range of materials into media for sensing these forces. Admittedly, as an experiment that involved a device for listening to an extraterrestrial signal, the work of Penzias and Wilson was by no means unique. There are comparisons here, for instance, with the experiments by Dame Jocelyn Bell Burnell (under the supervision of Anthony Hewish) with a four-acre radio telescope with which she detected the existence of pulsars. However, the Holmdel experiment differs, because unlike Bell Burnell's experiment, the source of the signal could not be identified with any specificity or clarity.

Penzias and Wilson's experiment is also a media experiment, however, because it poses the question of what it is to sense something whose signal comes from all directions at once. It is an experiment that foregrounds with particular acuity the question of whether the source of such a signal can be sounded in terms of an entity or object. Certainly, this kind of sounding might be understood as a technique disclosing entities whose scale and duration are beyond the scope of any human frame of reference. CMB radiation could be the signature, afterglow, or relic of what Timothy Morton calls a "hyperobject," an entity so vast and massively distributed that it cannot grasped within human frames of perceptual and conceptual reference.[57] CMB does not need to be grasped in terms of an entity, however. Indeed, one of the remarkable aspects of the experiment by Penzias and Wilson is that what is sensed in this case is not an entity but something excessive of this category, something subsisting between cosmic bodies, between planets, stars, and galaxies, something between worlds.

Rather than an entity, the Holmdel antenna sensed a kind of thermal background persisting as the afteraffect of a singular event. But it did not sense this thermal background as atmospheric, and for at least two reasons. First, the idea of the atmospheric suggests the presence of a medium in the absence of which affects and sounds will not propagate; CMB, in contrast, can be detected as a kind of background extra-atmospheric acoustic oscillation in matter. Second, the concept of the atmospheric is suggestive of a medium of variable intensity that can be backgrounded and foregrounded to different degrees. But CMB cannot be foregrounded. It remains in, or precisely as, background, without crossing a threshold such that it becomes palpable as an immersive medium.[58] Because of this, CMB is more properly understood in terms of the ambient than the atmospheric. The "ambient," as distinct from "ambience," which refers to the sensory qualities of spacetime from a human point of view, remains a diffuse, low-intensity background.[59] The ambient does not draw attention to itself: it is itself interstitial to attention. In the case of CMB, the ambient is heat and light, the afterglow of an event from which all entities emerge but which does not itself become an entity.

If it is excessive of the circumstantial specificity of terrestrial atmospheric things, the shape of this ambient background can still be understood in terms of an envelope, or surface, at least of a sort. This is a surface that is always receding from the observer, listener, or sensing device. The shape of the ambient is the shape of a period of recombination, taking place after the big bang, during which the universe cooled and electrons began to combine to form atoms, leaving photons to move freely. Peter Coles explains it thus: CMB radiation "appears to come from a spherical surface around the observer such that the radius of the shell is the distance each photon has travelled since it was last scattered at the epoch of recombination. This surface is what is called the last scattering surface."[60] Beyond this surface, everything is ionized in an opaque but bright fog. Coles offers another way of thinking of this surface, not as one of scattering but as what he calls a "surface of last screaming," the sound of which is the furthest in time and space from the observer that can be sensed.

Cosmologically, then, the significance of what Penzias and Wilson did was to provide evidence for the existence of the ambient heat and light of the cosmos, paving the way for confirmation of the expansionist theory of the universe. But the significance of this experiment as a media experiment is rather different: by converting ambient light and heat into sound—by sounding the ambient—they transformed the afterglow of an event into something that could be sensed as atmospheric. The atmospheric, in this context, became the threshold over which something ambient could be sensed rather than the signature of an entity. In the process, by sounding this signal through the familiar hiss of atmospheric radio static, the Holmdel experiment extended the limit of the envelope of media sensing beyond the arc of the gaseous atmosphere into the ambient cosmos. The experiment by Penzias and Wilson was, then, among other things, a media experiment in sounding, in sounding the limits of that which can be sensed as a signal. Following Michel Serres, we might call this signal the background noise of all media, the elemental media signature of a sonorous and luminous cosmos.[61] This signature is the ongoing refrain of an event whose afterlife might be grasped as the envelope of all envelopes.

ATMOSPHERIC SOUNDING FOR ELEMENTAL MEDIA

The balloon remains as a device for sounding the cosmological.[62] A series of balloon-borne sensing projects (TOCO, BOOMERaNG, and MAXIMA) begun in the late 1990s were used to detect what Bruce Bassett, Bob Nichol, and Daniel Eisenstein call "acoustic oscillations in the distribution of radiation and matter in the universe" as part of the project to map the Cosmic Microwave Background. These projects produced new maps of the sense-able universe defined not so much in terms of entities but rather in terms of gradients, oscillations, and density fluctuations.[63] In doing so, the flights of these stratospheric balloons drew together the kinds of practical experiments of Lally with the cosmological orientation of the experiments by Penzias and Wilson. They signaled the ongoing value of experiments that combined devices in the air and on the surface of the earth for trying to sound whatever Camille Flammarion had speculated might "reign in the interplanetary space in the midst of which worlds revolve."[64]

Such experiments are part of the ongoing process of sounding the volume(s) of spacetimes in ways that hold together different senses of media and medium: media as message transmission, media as medium through which messages move, and medium as vessel or container. To experiment with sounding is therefore to undertake a mixed-media experiment of sorts. These are mixed-media experiments that both reveal and stretch the atmospheric as a field of sensed variation in assemblages of devices and bodies of different kinds. In the case of the GHOST project this involves using the balloon as a vehicle with which to assay the qualities of the atmosphere as the medium through which this vehicle moves. In the case of

Project Echo it involves using the balloon as a surface against which radio signals can be reflected. In the case of the work by Penzias and Wilson, since developed further by experiments with stratospheric balloons and satellites, it involves sounding the signal of all signals, the cry of the cosmos as that which is always excessive of entities. Taken together, these are media experiments that disclose the very limits of the atmospheric as envelope of sensed elemental variation, the limits of what Frances Dyson calls the "reverberative atmosphere that surrounds, encloses, shapes, and sustains us."[65]

Sounding then, is an experiment in sensing variations and patterns in the atmosphere, and of foregrounding the condition of becoming atmospheric as an important threshold of that which can be sensed. Experiments with sounding point us toward a kind of extended media studies that is as much meteorological and cosmological as geological in orientation, a media studies that listens for refrains beyond the cry of the earth.[66] In some cases, this involves the use of various devices, including the balloon in its various incarnations, to pursue what Sasha Engelmann calls a "more-than-human" form of "affinitive listening" to the oscillations of the atmosphere as an envelope of sensible variation.[67] But this is also a kind of media studies whose attention is directed to the possibility of a sound or signal that requires no atmospheric medium, which in this case is the signature of the ambient light and heat between things. Listening or becoming attuned to these signals is not so much about the search for alien life but about rendering sonorous the extraterrestrial. Nevertheless, while moving beyond the geological to the cosmic, this kind of sounding is facilitated by a range of earthbound devices. Indeed, as part of what Michael Gallagher, Anja Kanngieser, and Jonathan Prior call an "expanded sense of listening," the earth itself might be becoming a device for such sounding, as we become more aware of the capacities of different materials from which it is composed to sense variations in previously unheard of particles.[68] As Jol Thomson and Sasha Engelmann suggest, these experiments involve using the material of the earth as a listening, sounding, or detecting device with which to pick up the traces of elemental particles that in any other circumstance could not be rendered present because they pass through things without mediating or being mediated by those things.[69] Here, the scale of sounding and listening goes far beyond the human body and the devices with which it is surrounded, and extends to massive devices designed to detect infinitely small perturbations in spacetimes.

All this suggests that it is necessary to develop a much wider conception of media studies, and of media geography, through which to grasp the circumstances under which the atmospheric comes to matter. The field of this sounding is far more expansive than the traditionally circumscribed spaces of acoustic phenomenology, and extends to the materialist question of how to pick up the traces of a vibrational milieu in excess of human capacities to sense.[70] Following thinkers such as John Durham Peters, we might think of this as a kind of elemental media studies concerned with exploring how far the relation between sensing and medium can be stretched via experiments with different devices.[71] This form of media studies would be concerned with, among other things, how forms of envelopment, whether the body, the balloon, the antenna, or the earth itself, provide sounding devices through which the atmospheric is both an earthly elemental condition for sensing and the threshold across which something more than earthly must pass before it can be sensed. This would be a media studies concerned with experimenting with sounding—in all aspects of that term—the limits of the envelope of possible sensing. Importantly, this limit is not necessarily defined in terms of the allure of an entity whose essence (as something unified) is to be withdrawn from us. Rather, it is the time-shifted limit of an envelope that never takes the form of an entity, but takes shape, however dimly, however faintly, as a shimmering, screaming surface always receding from whatever senses toward it.

7

When in the air the balloon draws together distinctive spaces and experiences of sensing. When on the ground, it affords opportunities for the generation of atmospheres that gather variously around the spectacle of ascension or, in other instances, around the promise of a form of affective immersion with and between other bodies. There is, however, another, and less heralded story to be told about how envelopment takes shape as the form of the balloon. This is a story about the work that goes into the material fabrication of envelopment. The "work" in this case consists of the combination of two related domains of expertise. The first is the repertoire of practices, involving varying degrees of technical complexity, that are required to allow an envelope to take to the air as an atmospheric thing. This repertoire includes everything from stitching to sealing to folding to checking for leaks. The second domain of expertise is more structural, and is organized around the question of how lighter-than-air things are sheltered in the period between their inflation and ascension. This domain of expertise involves the problem of how to engineer envelopes for housing other kinds of envelopes.

Why attend to these domains of expertise and the diverse experiences they involve? One reason is in order to foreground structures and practices operating in the boundary zone between ground and air. If being on the ground and being in the air are both characterized by relative stability (and by a strange sense of stillness in the case of the latter), the transition between ground and air is potentially more fraught: it is characterized by the passage between spacetimes of different consistencies, speeds, densities, and properties. Thinking about the work of envelopment, and the practices and structures of which it consists, might then be an opportunity to develop a vocabulary for grasping the transitional, but no less atmospheric zone, between ground and air in ways that do not juxtapose the concreteness of the former with the immateriality of the latter; instead, what might emerge is a mixed zone whose materiality is tensed instead by relative forces that are sensed in various ways.

The work and experience of envelopment is also differentiated across bodies. Clearly, one way in which this differentiation matters depends upon how far bodies can take to the air; the experience of being-in-the-air is different from that of being-on-the-ground. But it also matters because the mundane work of envelopment is distributed across different bodies on the basis of assumptions about the capacities of bodies to act in particular ways. The work of envelopment is striated by the experience of distinctive divisions of labor. Thinking about these divisions is a way of disclosing aspects of the untold histories of envelopment. It also encourages us to explore possibilities for experimenting with the work of envelopment in ways that might generate the conditions for ungrounding the practices and structures that shape what bodies can do as they take to the air.

HARD AND SOFT

As a technical process of fabrication, envelopment takes shape through a responsive relation between a material membrane and the elemental conditions of an atmospheric milieu. Envelopment, in these terms, is a process of partial enclosure generating a tensile relation between whatever is inside to the elemental variations of atmospheres outside. Even if it can be grasped in these terms, however, the process of envelopment is badly served by oppositional distinctions between the material and immaterial, between atmosphere and thing, or between inside and outside.

Atmospheric envelopes remain open to the milieu in which they move, responsive to changes in the conditions of these milieus. How then to grasp the real and sometimes palpable differences and variations that characterize forms of atmospheric envelopment—and conditions of exposure—in ways that do not posit sharp distinctions between these categories?

One way of doing this might be to think in terms of a gradient of continuous difference between what Michel Serres calls, rather enigmatically, and typically cryptically, "the hard and the soft."[1] While he uses them in different ways, for Serres these terms refer to an index of material differentiation, based on consistency, force, and durability, which cannot be mapped straightforwardly onto the distinction between the natural and the social. Serres uses these terms to think of the differences, notably, between infrastructural hardware and informational software, but also between skeleton and flesh, and between elemental categories like rock and air.[2] What he is trying to specify is not so much the difference between categories of entities but the difference between the kinds of relational forces that give shape to various forms. It is perhaps easiest to grasp this in terms of the relation between atmospheres and entities. On the face of it, an atmosphere is not "hard" in the same way, for instance, that a tree is. Its gaseous consistency is of a softer kind. And yet its movement might be considered hard: the force of the atmospheric variation we call wind can be hard or soft depending on the circumstances under which it affects or perturbs an entity. A breeze and a gale are both nothing more than moving air, but they possess very different capacities to act and to affect the entities they encounter.

Equally, the relation between an envelope and the elemental conditions of its atmospheric milieu can be hard or soft depending on whether this envelope is tethered to the ground—in which case it feels the hardness of the wind as a potentially destructive disagreement—or free to move, in which case the force of the elements is much softer. In that sense, the relation between hardness and softness is a relation of circumstantial variation in relative forces and speeds rather than a relation between discrete entities. And it matters only under certain circumstances of relative disagreement: circumstances found in that transitional zone between being on the ground and taking to the air.

As Serres's own writing demonstrates, thinking about the tensions between the hardness and softness of forms and forces is not just an occasion for speculating about a metaphysical problem. It is also an opportunity to explore how these tensions can be and are translated into the design, architecture, and fabrication of particular structures. And, simultaneously, it is an opportunity to explore how bodies of different kinds sense and become implicated in these tensions in different ways. In terms of the former, Serres's writing offers some orientation: he signals the importance of different "boxes" as envelopes for housing bodies. By this he means not only the volumetric envelopes of architectural shelters through which the exposure of bodies to the elements is modified—boxes for holding the world "at bay" and for softening the force of the elemental.[3] He also means the way in which such "boxes" are softened further by being dressed by other envelopes, in the form of carpets, curtains, plaster, decorations, and so on. And he also means the envelopes of sensation that soften the exposure of bodies to the hard forces of the elemental world, envelopes that mark the extent of the mingling of the human body with the affective turbulence of the world.

Serres is critical of how contemporary forms of life have enveloped us to the extent that we are no longer exposed to the elemental conditions of the world. This does not mean advocating a return to a harsh, romantic, primal exposure to the elements; it is a reminder that the envelopes we construct, whether architectural, representational, or social, tend to diminish the value of being exposed to the elemental. And it is also an invitation to experiment with practices through which bodies can learn to feel again the condition of this exposure. Serres's own writing reveals a joyous enthusiasm for a range of practices, from dancing to swimming to running, and the kinds of skillful involvement they demand. He has a particular fascination, born of various experiences in childhood and adulthood, with sailing. For Serres, sailing is a practice through which to cultivate an attunement to elemental agencies and forces via the capacities of a particular craft. And it is one that relies upon an envelope of sorts: the sail is a surface of exposure and partial envelopment through which atmospheric forces and variations are translated through the structure of a craft, its relation to another elemental mileiu—the water—and the skill of whoever is piloting the craft.[4]

Serres draws our attention to the value of modifying the exposure of bodies to their elemental circumstances. He affirms a lively, joyous, and at times muscular experimentalism. His arguments need a little tempering, however, not least because the opportunities for different bodies to experiment in this way, and the circumstances under which this experimentalism can take place, vary enormously. As Luce Irigaray's work reminds us, thinking of the relation between envelopes and bodies through terms such as "the hard" and "the soft" is complicated by the associations of those terms. Certainly, in terms of divisions of labor, hardness has tended to be associated with masculine work, and softness with feminine work. "Hard work" has been understood in terms of public work exposed to the elements, while "softness" is domestic, sheltered, and less energetic. As Irigaray puts it, in many domains, traditionally women have been left "the so-called minor arts: cooking knitting, embroidery, and sewing; and, in exceptional cases, poetry, painting, and music."[5] Men, in contrast, are tasked with what have been considered more muscular, more demanding, riskier roles. Attention to aspects of the history of the work of envelopment reveals some ways that this work reaffirms the distinction between hardness and softness. But it also has the capacity to point us to the circumstances under which it can be transformed as part of the experience of being-in-the-air, however modestly.

HOUSING BODIES

With the development of lighter-than-air travel in the late eighteenth century, the problem of envelopment was posed as an aerostatic one: that is, as a problem of how bodies could take to and be in the air. From the outset, however, this problem generated another: how to engineer forms of envelopment on the ground. When aloft the balloon moves largely in agreement with its atmospheric milieu. It moves horizontally with the movement and direction of the body of air in which it is enveloped, or responds vertically to the solar and terrestrial radiation to which it is exposed. On the ground, when tethered, the balloon is in disagreement with the elemental forces of the atmosphere. It moves at a different speed. It feels the force of atmospheric variations in ways that are harder than it does when aloft. This force becomes a potential threat from which the balloon envelope must be sheltered in a process of double envelopment. Before ever taking to the air, the balloon is suspended and tensed between envelopes and their outside, between the pull of buoyancy and the anchorage of a tether.[6]

From the very beginning the tensed relation between the balloon envelope and the atmospheric milieu to which it is exposed generated the requirement for another envelope. In 1784 Jean Baptiste Marie Charles Meusnier de la Place produced what was probably the very first design of a structure for housing lighter-than-air bodies.[7] The structure took the form of a tentlike canopy, tethered to the ground by guy ropes. Meusnier was a mathematician whose central interest was with the geometry of the curvature of surfaces. Elected to the French Academy, he was appointed to a commission to investigate prospects for lighter-than-air travel in the wake of the balloon flights by the Montgolfier brothers and Jacques Charles.[8] He also collaborated with Antoine Lavoisier on the analysis and production of hydrogen from water, and proposed a design for a "dirigible balloon of elongated ellipsoidal form." This design specified a hydrogen-filled gasbag (ballonet) housed inside a larger envelope filled with air. When air was added to the outer bag, the effective weight of the airship would increase and the airship could descend, with the process being reversed when air was removed. One version of his airship was to be 260 feet long and would carry a "crew of thirty and food for sixty days, able to circumnavigate the earth."[9] It was this dirigible that Meusnier's tentlike structure was designed to house during the periods when it was not in the air.

Neither Meusnier's airship nor his hangar was ever built; indeed, it was almost a century before large buildings for housing lighter-than-air bodies were constructed. And when they were, it was not in the shape of what, following Vilém Flusser, we might call the "wind walls" of Meusnier's tent. Rather, they took the form of "solid walls," fabricated from the materials of industrial modernity.[10] In 1879 the engineers Charles Renard and Arthur Krebs constructed a large balloon shed (or hangar) near Paris in order to house a dirigible airship. Constructed from metal and glass left over from the Paris Exposition of the previous year, what became known as Hangar Y was one of the first of a large number of similar structures constructed from the late nineteenth century into the early twentieth century.[11] The form of these hangars was defined primarily by the shape and volume of the bodies of gas they housed. Key here was the requirement to build structures without any pillars or struts. But the design of these structures was also shaped by the need to take into account the relations between the properties of these structures, the materials from which they were composed, the behaviors of airships, and the prevailing atmospheric circumstances and weather conditions at their location. A critical challenge posed by operating large airships in and around the structures designed to house them was the problem of how to position these airships in the direction of the wind. Large airships need to land facing into the wind, and crosswinds make it difficult to maneuver these craft into their hangars. One solution to this problem was to construct mobile hangers whose orientation could be adjusted according to the direction of the wind. For instance, Count Ferdinand von Zeppelin constructed a floating hangar on

Lake Constance at Friedrichshafen that could be turned in order to allow the airship to be housed as easily as possible. Such moving hangars were also constructed on land. The 135-meter-long hangar constructed by the Siemens-Schuckert Corporation at Berlin-Biesdorf in 1909 could be rotated along a specially designed track to accommodate landings during different wind conditions.[12] The design of these structures took into account the force and movement of the elements, the properties of gas-filled bodies, and the difficulty of balancing the relation between both.

Even when they could not rotate in this way larger airship hangars were designed to be responsive to their elemental conditions. Completed in 1929, the Airdock Hangar in Ohio was built by the Goodyear-Zeppelin Corporation as a facility within which to construct its airships. Proclaiming it the largest building in the world, *Popular Science* noted that its ends were rounded "to prevent the formation of disturbing air currents and sudden wind gusts" which might threaten airships when they were in the vicinity of the hangar.[13] The "most amazing feature of the design," the article continued, was "the one that permits the whole structure, with its acres of corrugated steel roofing, to stretch and change its size, to expand and contract with alterations in temperatures." A system of rollers and hinges allowed the hangar to stretch by over a foot, "warping back and forth" in response to the weather.[14]

Such structures lie somewhere between the hardness of traditional engineering constructions and the softness of certain kinds of inflatable architectural envelopes designed to be responsive to their atmospheric conditions. As architectural critic Reyner Banham wrote, such inflatable envelopes responded directly and continuously to microvariations inside or out, and "a blow directed at the enclosing skin would produce a flurry of reproachful quivering and creaking, quickly dying away as the even tenor of its normal breathing ways was resumed."[15] Clearly, structures designed to house airships don't go as far as inflatable architectures in terms of their dynamic responsiveness to the atmospheric conditions within and beyond their envelopes. At the same time, however, they complicate the stark division that Banham draws between hard (traditional) and soft (inflatable) architecture. In being ever so modestly mobile, in being able to rotate, to warp, to stretch, however minutely in some cases, these structures have the capacity to soften a little in response to the hardness of the elemental force of the atmosphere.

Paralleling the gasometer, these structures are also physical manifestations of the pneumatic pact of modernity: exoskeletons of the envelopes in which gas becomes the focus of various technical processes and operations. While the extent of these structures can be grasped in terms of their hard volumetrics, or in terms of calculable scale, they can also be grasped aesthetically, in terms of a voluminous atmospherics. Indeed, while considered and sometimes dismissed as works of engineering rather than architecture, their designers sometimes recognized their aesthetic qualities. Among the most famous were two hangars constructed at Orly, near Paris, in the early 1920s. Designed by the engineer Eugène Freyssinet, they were over one hundred meters wide at the base, and were constructed of "folded" parabolic ribs covered with a concrete membrane that created a surface of strength and rigidity in dynamic tension. Destroyed in 1944, these hangars are significant because they demonstrated the feasibility of prestressed concrete for producing spaces of enormous volume. They were also structures that generated particular aesthetic experiences. In an essay called "On the Sublime," Freyssinet wrote, "Anyone who goes into the hangar by the side entrances when the main doors are closed is overcome by a very powerful impression. . . . Even people predisposed to be hostile were not immune from it. The impression is due not just to the unusual size of the building; above all it is a feeling of equilibrium, harmony and order—an immediate sense of certainty that every detail is just as it should be."[16] For Freyssinet, the aesthetic qualities of the hangars emerged because engineering was never just a technical endeavor—it was also a process imbued with "passion and emotion."[17] But we might say that these qualities emerge also because even the most technical of envelopes has the capacity to generate an atmospherics excessive of measure and calculable volume, from which different conditions of experience can precipitate.

LIGHT WORK

If the elemental force of the atmospheric is translated through and into engineered structures, it is also felt in the bodies of those who work within and around these structures. This is the work of lightness: of making things lighter-than-air. To say that this work is light is not to suggest it is whimsical or inconsequential: the work of lightness is always a combination of hardness and softness, of the circumstantial tensing of relations of force.

There is an archive of this kind of light work: an archive of the work of fabricating, tending, and handling lighter-than-air things both inside and outside the protective envelope of their sheds and hangars. This work is circumscribed in turn by assumptions about which kinds of bodies can and should be engaged in hard work and soft work. A gendered division of labor here is stark: perhaps unsurprisingly, women, and traditionally seamstresses, have often been engaged in the processes of fabrication, checking, and repairing of fabric envelopes, with men, in contrast, often deployed to perform what is assumed to be the riskier work of handling and controlling inflated envelopes as they are moved outdoors and exposed to the elements. For instance, during the Siege of Paris by the Prussians in 1871, balloon factories were established in the Gare du Nord and the Gare d'Orléans. As James Glaisher describes it, the collective process of fabricating these envelopes and the different stages of this process "presented an extraordinary scene, the Orleans Station in particular. . . . This operation, which required accuracy, was performed by numerous workwomen, under the personal direction of M. Godard; there might be seen every day nearly a hundred women, silent and attentive, marking with mathematical precision, by means of a pin and card, the distance between each point."[18] While women did much of the work in the factory, men—chiefly sailors— undertook other aspects of the fabrication process. Glaisher writes, "Sailors seemed to be quite at home with their work, painting, varnishing, weaving nets, twisting cables, and finally taking charge of the balloon on its journey."[19] Irigaray's critique of envelopment should sensitize us here to the implicit alignment of the softness of working with fabric envelopes, of female sensibility, and of the condition of working inside. A drawing of the Lachambre balloon factory in Paris made by Albert Tissandier in 1883 (figure 7.1) depicts this division of labor. In the drawing, seamstresses are working, surrounded not only by balloon envelopes but also by inflated shapes of all kinds: a giraffe, a tiger, a crocodile, and comical human figures. In such a scene, women undertake the work of fabricating envelopes used for aerial voyages undertaken almost exclusively by men.

World War II provides further evidence of this division of labor, but also, importantly, the circumstances under which it could be challenged, and in ways that question assumptions about the relations between bodies, envelopment, and exposure. In the United Kingdom, Balloon Command deployed significant numbers of the Women's Auxiliary Air Force (WAAF), whose work was documented by Ministry of Information photographers and artists. Notable fashion photographers such as Cecil Beaton captured a visual record of women carefully applying varnish to the fabric of balloon envelopes, or, clad in overalls and without shoes, working inside

FIGURE 7.1 Henri Lachambre's Paris balloon factory, 24 passage des Favorites, Paris, Vaugirard, August 1883. Drawing by Albert Tissandier. Graphite, ink, and lead white. Library of Congress, Wikimedia Commons.

partially inflated envelopes tending to or repairing the fabric.[20] War artists including Leslie Cole (figure 7.2), Robert Sargent Austin, and Laura Knight rendered similar scenes, depicting women working inside or outside the balloon envelope in various stages of inflation and deflation.[21] In such scenes, women perform acts of careful tending and repair: the remit of their relation to envelopment confined to the soft, enclosed, intimate work of being inside.

The kinds of gendered associations that informed the work of war photographers and artists are illuminated further in some of the photographs that Cecil Beaton took before the war. In 1930 Beaton photographed the three women of "The Soapsuds Group" (Baba Beaton, Wanda Baillie-Hamilton, and Lady Bridget Poulett) surrounded by white balloons and sheets of the

FIGURE 7.2 Leslie Cole, *Two WAAF Balloon Workers*, 1941. Watercolor on paper. H 400mm Width 575, Imperial War Museum.

newly invented translucent cellophane (figure 7.3). In other photos, Beaton staged portraits of "bright young things," including the American actress Tallulah Bankhead, similarly draped with, surrounded by, and to some extent enveloped by balloons (figure 7.4). In such scenes, as Laura Levin has suggested, balloons, new synthetic fabrics, and bodies were staged in order to compose a commodified aesthetic arrangement through which to create a soft, luminous atmosphere accentuated by the reflective sheen of alluring surfaces.[22] The bodies of these women are bathed and partially enveloped in this atmosphere in ways that suggest a kind of seductive association between the softness of the form of the balloon and the female body. Crucially, in these images the relation between bodies and materials is essentially static: both are immobilized as objects under the scrutiny of a gendered photographic gaze.

World War II did however generate conditions under which the relation between bodies, the work of envelopment, and experiments with atmospheric things was unsettled. The particular demands of total war opened

FIGURE 7.3 "The Soapsuds Group" at the Living Posters Ball, 1930.
Photo by Cecil Beaton.

up the strict gendered division between a kind of nimble-fingered precision and a more physical form of corporeal exertion, between the
softness of being enveloped and the hardness of exposure to elemental
conditions. Women took on roles often assumed traditionally to be the
preserve of men, including the control and handing of balloons exposed
to the elements. A painting by Laura Knight—*A Balloon Site, Coventry*
(1943)—depicts this kind of situation (figure 7.5). In the painting, four

FIGURE 7.4
Tallulah Bankhead,
n.d. Photo by Cecil
Beaton.

FIGURE 7.5
Laura Knight,
*A Balloon Site,
Coventry*, 1943.
Oil on canvas.
H 1242mm
W 1495 mm. Imperial
War Museum.

women of the Women's Auxiliary Air Force, supervised by another, are pulling on a rope to which a large barrage balloon is attached.

The handling of the balloon in these circumstances could be and often was a strenuous activity. It required strength and expertise, in addition to knowledge of the behavior of balloons under different weather conditions. Balloons needed to be positioned and adjusted, ropes needed tightening and loosening, and knots needed to be checked. Recounting her experience, one veteran put it thus:

> We all had ugly hands, scarred by rope burns and ingrained with oil from handling pulleys and shackles. There were idyllic summer days when the barrage floated 5,000ft above us and we lounged in the sun splicing new guy ropes, but also there were stormy days and nights when we had little rest as we struggled to keep our balloon on its storm mooring. To achieve this the bow must always be kept into the wind, not easy in a gusting, veering wind. Each guy rope had to be moved in turn, one point at a time, whilst another girl dragged its 56lb concrete ballast block with it. As the balloon pitched and tossed in the wind the guy ropes sprang taut and the ballast blocks swung into the air, just at the right height to smash a kneecap, if you did not stay alert. Finally the great silver hulk was edged back into wind, perhaps only to repeat the process one hour later when the direction of the wind changed yet again. To be woken by the duty picket shouting "Out of wind!" was not a welcome sound. It was also important that the balloon did not rip her belly on her concrete bed or on her own wire rigging, because she was only made of Egyptian cotton covered in silver dope. To this end, we crawled on hands and knees to adjust the heavy canvas ground sheets under her. It was important too that we did not get entangled in the shifting wire rigging.[23]

Knight was especially attentive to scenes of corporeal effort and expertise performed by women at work. During the early and mid-twentieth century she had been a key figure in the emergence and growing visibility of the female artist as a legitimate public figure in Britain. Moreover, she had a particular interest in athletic, skillful forms of expressive embodiment; among the many subjects she painted were boxers, circus performers, clowns, acrobats, aerialists, and ballet dancers.[24] Of an encounter with one circus aerialist, she wrote, "When walking along the Brixton Road one Saturday afternoon with an elderly circus artiste, he stopped me to say: 'Time was when I'd watch such a crowd plodding along, and say to myself, "you walk the ground—I walk the air.""[25] Even if they are not engaged in the kind of aesthetic aerialism of the circus artist, there is something about the work of the WAAF women in *A Balloon Site* that echoes the forms of skillful embodiment depicted in Knight's other work.

When considered in the context of responses to its performance the significance of WAAF work is heightened. One newspaper report at the time put it thus: "Most of the balloon sites are in exposed places, living conditions are very rough, and to handle a balloon in a breeze is heavy work even for a physically fit man. An unfit man would be entirely incapable of doing the work. It is definitely a man's job, and the men engaged on it have no reason to be self-conscious about their tasks."[26] As Tessa Stone has shown, even in the context of the shifting meaning and materiality of work under conditions of total war, when women were deployed to do "dangerous" roles, the description of these roles sometimes became downgraded in order to make it easier for men to come to terms with the active participation of women. Nevertheless, in performing this work, the women of Balloon Command were modifying, however modestly, the relations between envelopment and exposure, between bodies working inside and outside, and between bodies suspended between air and ground. In the process, the distinction between the hard work of exposure and the soft work of envelopment was ungrounded just enough to make a difference. This is not to say that these minor experiences have the capacity to transform wider patterns of relations between bodies; rather, they are significant in that they point to the circumstances under which the tensed relations between envelopment and atmosphere can be refigured, however modestly.[27]

CHOREOGRAPHIC LIGHT WORK

The work of envelopment in the transitional zone between ground and air also involves particular experiences of being aloft: experiences that are partially tethered but that also generate senses of being partially unmoored. This dimension of the work of envelopment might be understood as a form of choreography. As William Forsythe reminds us, choreography is about the organized arrangement of bodies, objects, and environments in ways that offer opportunities for stretching the envelope of practiced movement.[28] While the work of choreography is practiced and performed in obviously aesthetic contexts, opportunities for minor choreographic experiment can be found in a range of other circumstances. This includes the work of fabricating, tending, and launching lighter-than-air things, involving the arrangement of relations between bodies, cables, fabrics, and elements in the production of an envelope that can become lighter-than-air.

Even if much of this work takes place on the ground, as it were, it sometimes also involves a modest form of aerialism suspended in the transitional zone between air and ground. Consider again the Andrée expedition, the exposed launch site of which was located in the northwestern part of the Svalbard Archipelago. This location, and its exposure to the elements, necessitated the construction of a house within which the balloon could be inflated and sheltered in advance of its launch. Housed thus, the balloon became stable enough to act as a reasonably firm surface over which members of the expedition could walk. Pictures reproduced in balloon-maker Henri Lachambre's account of the expedition show groups of sailors on the meshed dome of the envelope, testing it for "gas-tightness," with only a minimal amount of cables and pulleys for support (figure 7.6). Lachambre described the process thus:

> Eight and sometimes ten of us were at work on the dome of the inflated balloon, and we had to perform compulsory gymnastic feats in order to support ourselves amidst the cordage of the net. The sailors, being accustomed to this kind of exercise, climbed about the balloon quite at their ease; but I must confess that at first I had a slight feeling of dizziness; this, however, soon passed off. It was a curious sight to see so many men on this silken envelope, which is the only barrier to the gas. The fact is unprecedented in the history of balloons.[29]

There are some obvious echoes between this kind of work and other activities. It echoes, for instance, the forms of agility and expertise performed by sailors as they clamber about on the rigging of sailing ships high above moving decks. It echoes the skilled, bodily dexterity performed as part of the construction of engineering structures, from the modest to the monumental.[30] And, perhaps more faintly, it echoes the work of more obviously performative aerialists, from popular theater entertainers in the nineteenth century to more contemporary wire-walkers.[31] Indeed, the mode of surface walking performed by Andrée's sailors can be understood as a modified version of wire walking: the surface of the envelope, taut and tensed like a wire, becomes, in effect, an elastic spacetime across which to perform a style of practiced yet modestly inventive mobility. As Steven Connor writes, the wire walker transforms the space of the wire: "The wire-walker aims to occupy rather than merely to penetrate space, to tangle up the line into a maze, to thicken the infinitesimally thin itinerary of the

FIGURE 7.6 Sailors working on the upper surface of Andrée's balloon, Spitzbergen, 1897. Photo by Nils Strindberg. Grenna Museum—Andréexpeditionen Polarcenter.

wire into a habitat."[32] Similarly, if not so deliberately, the act of walking across the surface of an envelope gives sensorial depth to a volume. Along the shaped surface of the envelope, bodies generate spacetimes of different volumes and affective vectors.

In his silent short film *The Balloonatic* (1923), Buster Keaton transforms the surface of the balloon envelope, however momentarily, into a comedic spacetime.[33] Early on in the film Keaton encounters a tethered balloon surrounded by a small group of serious-looking gentlemen and military officers. At their request, and with the aid of a ladder, Keaton climbs on top of the envelope, apparently to make some minor repairs and adjustments. At this point the balloon is unexpectedly untethered, and, without realizing it at first, Keaton finds himself adrift alone above the city. The upper surface of the envelope briefly becomes a stage upon which Keaton performs, the passage ending when, after climbing down into the basket, he ends up shooting a hole in the envelope; predictably, both the balloon and Keaton then crash to Earth in a rural setting.

FIGURE 7.7 Still from *The Balloonatic*, starring Buster Keaton, directed by Edward Cline and Buster Keaton (Buster Keaton Productions, 1923).

Envelopment also involves other forms of light work, although not so comedic. The scale and qualities of the envelope mean that it is not always possible to walk on its surface in order to check it or repair it. In such cases, other kinds of choreographies become necessary; these involve the careful arrangement of tethered bodies and envelopes on the ground and in the air. We can see this if we return to Project Echo. The June 21, 1963, issue of *Life Magazine* featured an image of a balloon, or, more accurately, of two balloons, the larger of which was one of those used during Project Echo (see chapter 6).[34] Under the title "A Big Echo's Little Pal," the text that described the image ran as follows: "In a hangar at Lakehurst, NJ, a little balloon, tethered to the ground, floated up and down beside a gigantic balloon while its passenger looked the big bag over for possible flaws. A technician, he was inspecting Echo II, a 135-foot-tall satellite which will be orbited later this year. Like the smaller Echo I, which is still orbiting,

FIGURE 7.8 "A Big Echo's Little Pal," *Life Magazine*, June 21, 1963.

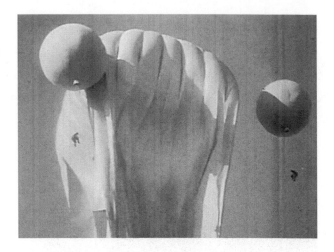

FIGURE 7.9 Still from *The Mirror*, directed by Andrei Tarkovsky (Soviet Union, Mosfilm Productions, 1975).

it will be used to bounce radio signals back to earth in communication tests."[35] It is as if the smaller balloon is in orbit around the larger balloon to which it attends, while its position and height are controlled from a tether below. This is a different kind of light work, in which envelopes become tenders for moving around other envelopes, in the choreographed operation of a form of orbital attention.

This way of tending to lighter-than-air things is not unique. Something similar is depicted in one of the many remarkable sequences in Andrei Tarkovsky's film *The Mirror* (1975). The footage is taken from a longer film of the launch in 1933 of USSR-1, a stratospheric balloon that reached an altitude of nineteen thousand meters during a flight that lasted about eight hours. The commentary to the Pathé film of the launch from which Tarkovsky's scene is drawn states that a "novel feature of the ascent was the two small captive balloons, each carrying one man, which hovered around the stratospheric balloon while it was on the ground, to help raise and lower ropes and inspect possible weak spots in the envelope, 112 feet in diameter."[36]

The presence of these balloons in *The Mirror* echoes Tarkovsky's use of this device in some of his other films. Notably, in *Andrei Rublev* (1966), for instance, a medieval monk takes to the air in a rudimentary balloon. And in *Solaris* (1976), "prints of early balloons adorn the walls of the dacha (whose hero is an astronaut)," while in the opening scenes of the same film a small yellow balloon is seen caught in a tree.[37] These scenes illustrate the importance of themes of lightness and levitation in Tarkovsky's films more generally. The stratospheric balloon scene in *The Mirror* accentuates this. After the small tender balloons return to Earth, the larger balloon ascends into the sky in a moment that draws together the allure of ascension and the affects of release. And yet the scene is also important because it reveals something of the softness of the light work required to allow things to become lighter-than-air: it reveals the kind of aerial choreography involved in tending, checking, rigging, tethering, and enveloping that makes the process and experience of ascension possible. This form of work does not necessarily unsettle gendered divisions of labor around envelopment. An event of hard, heroic ascension remains at the center of the film. But the forms of work that orbit this event are far softer, lighter, and much less adventurous. They gesture to possibilities for complicating, albeit modestly, the tension between the hard and the soft through practices and spaces of envelopment.

The work of envelopment in the contexts outlined above is circumscribed by assumptions about what kinds of bodies can be exposed to different conditions. But even under such conditions opportunities for a minor form of circumstance-specific experimentalism can emerge. In the cases above, experimentalism is an accidental by-product of something else. Equally, there is a risk of over-aestheticizing the experience of working with and within envelopes of different kinds. But the modest archive of these envelopes and the experience of working with them nevertheless offers reminders of how specific practices and structures temper the relation between envelopment, exposure, and what Serres calls the hard and the soft. They move us to think about other possibilities for choreographing arrangements between envelopment, bodies, and the elemental force of atmospheres. And to think about how these arrangements might allow bodies to become exposed, in minor ways, to the condition of being-in-the-air.

BEING IN ORBIT

It is early December 2013 and I am walking across a series of tensed wire meshes high under the roof of the atrium of the K21 Gallery in Düsseldorf. Between the meshes are a number of large inflated spheres, some translucent, and some with reflective, metalized surfaces. The presence and position of these spheres deforms the meshes, creating a tensile space composed of different gradients and inclines. The cable mesh is taught, flexible, and dynamic. It moves when something moves along it. As I move, I generate vibrations that travel along and across the mesh, and I can also feel those vibrations generated by others moving elsewhere along the structure. At times the entire work seems to be in motion, however gently.[38] To move around the spheres that deform and stretch the mesh is to recall other senses and stories of being aloft. It is to be reminded of the pneumatic kinetics experienced on the surface of Donald Barthelme's fictional balloon above Manhattan, a surface that was "so structured that a 'landscape' was presented, small valleys as well as slight knolls, or mounds; once atop the balloon, a stroll was possible, or even a trip, from one place to another. There was pleasure in being able to run down an incline, then up the opposing slope, both gently graded, or in making a leap from one side to the other. Bouncing was possible, because of the pneumaticity of the surface, and even falling, if that was your wish."[39]

Installed at the K21 Gallery from 2013 until late 2015, *In Orbit* is a work by the Argentinian-born and Berlin-based artist and architect Tomás Saraceno. This elaborately engineered installation consists of a number of connected wire meshes strung in tension below the roof and about twenty-five meters over the atrium of the gallery space. *In Orbit* takes up some of the themes and materials with which Saraceno has experimented in other work, not least the relation between inflation, envelopment, and the production of relational surfaces and forms.[40] For Saraceno, the aim of such works is to generate a kind of relational sensibility attuned to the entanglement of bodies and the environments in which they are enveloped and move.[41] Indeed, perhaps more so than any other contemporary artist, Saraceno's work foregrounds and choreographs the generative tension between forms of fabricated envelopment and the demands of exposure to the elemental conditions that sustain these forms.[42] Others, most notably Sasha Engelmann, are exploring the complexities and promise of this work for thinking through a range of conceptual, aesthetic, and ethicopolitical issues.[43] While taking some orientation from such work, my aim in thinking with it here is more limited: I am interested in how *In Orbit* provides an occasion to experiment with the kinds of experiences of the transitional zone between ground and air that characterize the archive of light work outlined above. That is not to say that it transcends all the problems with such activities, but it certainly points to the conditions under which they might be modified.

In Orbit is, among other things, an experiment with different forms, surfaces, and experiences of envelopment. In the shapes of *In Orbit*, the arrangement of spheres and meshes generates surfaces and volumes of partial envelopment over, across, and around which bodies can move and be moved. The effect is a kind of relational-topological surface of elastic

FIGURE 7.10 *In Orbit*, Tomás Saraceno, installed at the K21 Gallery, Düsseldorf, December 2013. Photo by author.

volumes composed of moving shapes and forms in tension.[44] To enter *In Orbit* is to enter a work where being inside and being outside are variations along a surface of experience rather than neatly demarcated space-times. To enter *In Orbit* is to negotiate a work in which being-in-the-air is at the same time to experience a different kind of groundedness along a vibrational surface. This work does not so much offer a straightforward experience of being aloft. Instead, it offers an experience of voluminous envelopment in which the distinction between ground and air is no longer clear-cut, and where being-in-the-air might also be facilitated by surfaces of envelopment that are supportive without feeling like grounds as such.

In Orbit stretches other distinctions, one of which is the distinction between the category of entity and its excess. Of course, a work like this is an entity. Equally, the spheres are entities, and the mesh is an entity. And yet, if we are to think of what is being choreographed here, it is not so much entities, bodies, or the relations between these entities. Rather, *In Orbit* is a work that choreographs forces in tension in ways that modify the tensed spacetime between atmospheric envelopment and exposure. It is a tensed prehensive structure whose variations are felt differentially in the bodies of those who walk across and within it.

And yet, when compared to the soft, light elementalism of other works by Saraceno, particularly *On Space Time Foam* (2012), *In Orbit* seems harder, more industrial, and more obviously engineered, a sense amplified by the fact that visitors are required to wear overalls and special shoes in order to explore the work. But it is this industrial dimension that allows *In Orbit* to echo elements of the archive of the light work of envelopment sketched above. On one level, *In Orbit* does this through its structure. The uncanny similarity between the structure of *In Orbit*, and of Meusnier's design for an airship hangar, remind us that the distinctive kinds of structures used to house lighter-than-air bodies—hangars, sheds, and tents—incorporate a certain softness that is responsive to the hardness of the elements. In some cases they involve the design of mobile hangars that can rotate in response to the wind, facilitating the choreography of bodies in the air and on the ground. *In Orbit* does not rotate in this way, of course. But it is a structure for enveloping envelopes in order to facilitate and modify their exposure to an elemental outside. The shape of an inflated envelope deforms a taught mesh or fabric along a surface of curvature, working to hold in tension the forces of hardness and softness. In the case of *In Orbit* there is the hardness of the gallery building itself, of the anchor points and the materials from which the building and the mesh are constructed. And there is the variable hardness and softness of the bodies that move around it, defined not so much by their form but by the degree of tension they experience in relation to the surfaces along which they move.

In Orbit also echoes, however unintentionally, aspects of the archive of embodied experience through which the shape and movement of atmospheric envelopes were choreographed in earlier theaters of operation, including balloon operations during World War II and Andrée's polar expedition. The most obvious sign of this is the distinctive, if unremarkable, mode of attire required to enter *In Orbit*; overalls and shoes with grip are required, both of which are provided by the attendants in the gallery. It is easy to imagine uniformed figures at work in airship hangars, tending and repairing barrage balloons. And it is easy to imagine, and indeed experience, the process of becoming accustomed to the sense of being ungrounded. Tentatively at first, movement on this mesh unfolds as the emergence of a form of sure-footedness that never quite takes itself for granted, tensed as it is as an ongoing responsiveness to the moving, shifting surface of the envelope. To move here is to experiment with the

FIGURE 7.11 Tomás Saraceno, *In Orbit*, K21 Gallery, Düsseldorf, December 2013. Photo by author.

weight of the experience of being aloft while supported by little more than a mesh of tensile cable. To move here is to sense something of the vertiginous, at least to begin with, as bodies get used to being on a surface that is simultaneously moving, supportive, and full of holes. This form of work is not immediately or eventually joyous. And it is differentiated at least in part on the basis of the different capacities and trajectories of the bodies that move across a surface tensed between points and around spheres.

UNGROUNDING

Luce Irigaray writes that the "transition to a new age requires a change in our perception and conception of *space-time*, the *inhabiting of places*, and of *containers*, or *envelopes of identity*. It assumes and entails an evolution or a transformation of forms, of the relations of *matter* and *form* and of the interval between.[45] Irigaray's aim in making this claim is to challenge the way in which woman has traditionally fallen between the category of envelope and thing: to imagine and invent a new place for woman. To claim that *In Orbit* deliberately pursues a similar aim would be a mistake. And yet it is an architectural experiment that reminds us of the conditions under which the relations between bodies and envelopes can be stretched. On one level, it reminds us of how the process of working with lighter-than-air things can unsettle long-standing divisions of embodied labor and experience. But it also produces surfaces and spheres in which the condition of being enveloped is reworked in ways that might encourage new modes of tentative relation with the elemental forces to which bodies are differentially exposed.

In Orbit is a reminder that experimenting with the relation between envelopment and exposure can involve the choreographic arrangement of forces in tension in the transitional zone between being-on-the-ground and being-in-the-air. This is an arrangement that has relatively fixed points of attachment and structural security, but only in order to open up the relations between bodies to creative variation. This is a form of choreographic arrangement that while providing some kind of constraint does not necessarily involve determining how bodies move. Rather, *In Orbit* is a "choreographic object" in the sense that William Forsythe uses that term because it offers an alternative "site for the understanding of potential instigation and organization of action to reside."[46] Such highly engineered choreographic arrangements can expose bodies to forces outside their immediate envelope of experience in ways that provide new possibilities for moving. Such experiments will never radically redraw the terms of relation between bodies, ground, and atmosphere in every circumstance. However, *In Orbit* reminds us of the value of experimental envelopes for holding in tension the force of the hard and the soft in order to generate possibilities for moving differently. The value of such experiments is, at least in part, how they generate envelopes supportive enough to expose us, however briefly, to the tentative groundlessness of our atmospheric being and becoming.

8

One of the weirdest British TV series of the 1960s, and perhaps one of the weirdest TV series ever, is *The Prisoner*.[1] Developed by and starring Patrick McGoohan in the lead role, the series is set in a fictional rural seaside town whose inhabitants, identified only by number, are all under constant surveillance. The administrator of the town is a character called No. 2, who acts, in turn, as the visible source of authority for a shadowy figure called, unsurprisingly, No. 1. One of the devices used to keep residents in check is a large white weather balloon called *Rover*, whose appearance and operation is often accompanied by a sinister roaring sound. Even if its real function is never fully explained in the series, *Rover* can coerce and incapacitate inhabitants, enveloping them fully to the point of death by suffocation if they make an attempt to escape or if they deviate from the rules. In one scene from the first episode, the inhabitants of the village are commanded by No. 2, via a loud hailer, to freeze, or "be still." Everyone in the central piazza of the town obeys, with the exception of

FIGURE 8.1 Still from *The Prisoner*, created by Patrick McGoohan, United Kingdom, ITC Entertainment, 1967.

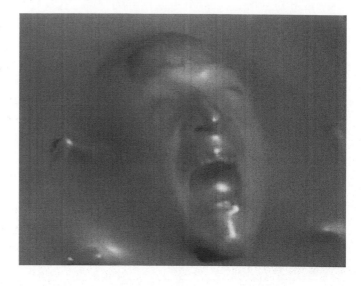

one man. Quickly, *Rover* appears and launches itself at this individual. The action then cuts to a remarkable image: the face of the man, his mouth agape in the shape of a silent scream, pressed in horror against a translucent latex membrane (figure 8.1). McGoohan's character (identified as No. 6 in the series) then turns to No. 2, asking, "What was that?" No. 2 replies, "That would be telling." In another scene, McGoohan's character tries to escape from the town in a speedboat but, after falling overboard, is chased and enveloped by *Rover*, controlled remotely from a well-staffed operations room.

Viewed now, *The Prisoner* seems an uncannily prescient vision of the worlds of surveillance, monitoring, and targeting in which many of us feel we are enveloped, worlds from which it seems increasingly difficult to escape. These are worlds, as Peter Adey has argued, defined by the generation and modulation of particular kinds of security atmospheres.[2] While the security of the world in which the character of *The Prisoner* is trapped is not reducible to aerial devices and technologies, it is nevertheless fair to say that aerial assemblages are critical to the infrastructures through which these security atmospheres are monitored, modulated, and targeted. As scholars such as Adey, Derek Gregory, Caren Kaplan, and Paul Virilio have shown, there is a long and complex history of the emergence of these systems of surveillance, targeting, and violence, through which geopower is performed and rendered operational via the capacity to maintain airborne and atmospheric presence through an aeromobile gaze.[3]

These aerial assemblages of atmospheric security are enabled by at least two promises, both of which are dramatized by *The Prisoner*. The first is dirigibility, or the susceptibility of devices and bodies to some kind of operational control, proximate or remote. *Rover* embodies and enacts this promise in ways that anticipate the more recent emergence of the drone as one of the key technical objects of contemporary forms of aerial life. In its more sinister deployments, the drone promises to render the lower atmosphere into a zone of continuous dirigible movement at a range of scales, a zone in which aerial devices can track and trace errant trajectories and, if necessary, act upon or against them according to a series of rehearsed and routinized protocols.[4] The second promise dramatized by the figure of *Rover* in *The Prisoner* is of continuous persistent presence as a necessary condition for generating a protective envelope of sensing. Here the figure of the drone (or *Rover*) can be understood in relation to a wider set of devices and systems that work to maintain envelopes of what Ian Shaw calls "full spectrum" sensing.[5]

While not deployed as widely as the drone, nor offering as captivating a critical target, the balloon, both as a tethered aerostat and as a dirigible, continues to participate in systems that, sometimes in tandem with drones, sustain atmospheres of security through envelopes of surveillant sensing. Such tethered systems have been deployed recently in theaters of military operation, including Afghanistan and Iraq, to complement other geopolitical assemblages. They have also been deployed in the United States. In 2014 the US military began testing one of these systems in the vicinity of Washington. Developed by Raytheon, the system is called the Joint Land Attack Cruise Missile Defense Elevated Netted Sensor System, or JLENS for short. According to Raytheon, the need for JLENS arises because "Airplanes, drones and cruise missiles pose a significant threat to people, population centers, key infrastructure and our military." Each JLENS system, which is called an "Orbit," consists of two tethered aerostats "that float 10,000 feet in the air. The helium-filled aerostats, each nearly as long as a football field, carry powerful radars that can protect a territory roughly the size of Texas from airborne threats."[6]

Costing $2.7 billion, the system was heralded for its capacity to track objects up to 340 miles away and to stay aloft for thirty days at a fraction of the cost of more conventional surveillance systems. Central to the operational logics of the system is the promise of persistent presence and continuous sensing. In reality, however, it has proved far less effective than its publicity material suggests, with estimates of its actual endurance

FIGURE 8.2 Schematic diagram of Raytheon's JLENS surveillance system, NORAD and NORTHCOM Public Affairs, 2014.

being measured in days rather than weeks. Equally, it has been unable to prevent the materialization of the threats it was supposed to be able to identify. On April 15, 2015, a Florida postal worker flew a small gyrocopter onto the West Lawn of the US Capitol, undetected by the JLENS system.[7] Then, in late October 2015, one of the blimps became untethered and broke free from its base at the Aberdeen Proving Ground in Maryland, drifting one hundred miles across Pennsylvania.[8] As it drifted it dragged cables that damaged power lines and disrupted electricity distribution to approximately 35,000 people.

The untethering of one of the JLENS aerostats is a reminder that the infrastructural assemblages that become the objects of so much critical interrogation from within the contemporary social sciences and humanities are sometimes a lot less effective, and a lot more wayward, than often assumed. At the same time, this untethering provides an opportune occasion for thinking about how the politics of atmospheric envelopes sometimes takes shape through processes other than dirigibility and persistent presence. In becoming untethered, the JLENS aerostat complicates the promise of control by becoming susceptible to the directional vagaries of the wind. At the same time, by becoming an object that drifts according to the wind, the JLENS aerostat generates its own kind of problematic situation: an emergency of sorts that disrupts the calculable volumes of regulated airspace. In doing so it reminds us how the capacity of an object to drift with the wind can, in certain circumstances, generate a situation that demands an urgent response whose effectiveness is complicated by the waywardness of the object in question (and its elemental milieu).

In this case such untethering is unintentional. However, there is a minor history of deliberate experiments that have used the capacity of the balloon to drift with the movement and trajectories of the atmosphere in order to distribute objects, ideas, affects, and in some case weapons.[9] In many instances these experiments have a malign intent, insofar as they aim to unsettle, disrupt, or puncture atmospheric envelopes of security by operationalizing meteorological trajectories and variations. Attention to some of these experiments, however, can also encourage us to speculate about the possibility that releasing simple technical devices to drift with the wind might be cautiously affirmed as a political act. In certain circumstances it may well be important to disrupt security atmospheres. Either way, the act of releasing something in order to drift with the wind might be understood in terms of the performance of an atmospheric mode of address: this is a form of address that complicates the political logics of location not so much to target a population from the air but to do atmospheric things that might call forth more benign collectives and publics.

"A RAIN OF FIRE BOMBS"

In 1670 an Italian Jesuit, Francesco Lana de Terzi, produced a design for a flying machine based upon four evacuated copper spheres. Lana de Terzi's theory was that the difference between the density of the sealed copper spheres and the surrounding air would allow the craft to fly. It would never have worked, not least because the spheres would have collapsed on themselves unless they were made from a material too heavy to allow the craft to fly. Lana de Terzi was confident, however, that any practical difficulties in designing this ship could be overcome in time. But he still thought it would never fly because God would not allow a craft of such destructive potential to take to the air. After all, as Lana de Terzi wrote, this was a machine that "would create many disturbances in the civil and political governments of mankind." He concluded his discussion of the aerial ship with a prescient vision:

> Where is the man who can fail to see that no city would be proof against surprise, as the ship could at any time be steered over its squares, or

even over the courtyards of dwelling-houses, and brought to earth for the landing of its crew? And in the case of ships that sail the seas, by allowing the aerial ship to descend from the high air to the level of their sails, their cordage could be cut; or even without descending so low iron weights could be hurled to wreck the ships and kill their crews, or they could be set on fire by fireballs and bombs; not ships alone, but houses, fortresses, and cities could be thus destroyed, with the certainty that the airship could come to no harm as the missiles could be hurled from a vast height.[10]

Lana de Terzi's vision was of a dirigible flying machine that could be steered toward a particular point or target. Practical versions of this craft were not realized until World War I, but other, earlier experiments sought to operationalize winds as infrastructures of distribution. While besieging Venice in 1849, the Austrian army was prevented by the shallow waters of the lagoon from sailing its ships and heavy cannon within range of the city. This situation provided an opportunity for an Austrian lieutenant, Franz von Uchatius, to deploy a balloon bomb made from varnished linen and paper, of which about two hundred were launched. Even if the physical or psychological impacts of the experiment remain unclear, the episode successfully demonstrated the practical possibility of balloons as weapons of war.[11]

As the range of guns increased during the late nineteenth century, the kinds of balloon bombs used in Venice were rendered redundant as accurate weapons. The more general difficulty with using balloons as strategic weapons, notes Charles Ziegler, was that they lacked the precision required to avoid killing noncombatants. Balloons were not used again for bombing during the nineteenth century, and a five-year ban on their deployment was ratified at the 1899 Hague Conference. Such restraint was tested during World War I: the possibility of launching indiscriminate retaliatory strikes became more palatable as this conflict dragged on.[12] In the United States, experiments with balloon bombing were undertaken under the authority of physicist and future Nobel laureate Robert Millikan, head of the National Research Council, who would use balloons to confirm the existence of cosmic rays during the interwar period. Notwithstanding claims about the dishonorable nature of this weapon, there was considerable support among Millikan's team for the potential of balloons to generate a destructive and morale-damaging "rain of fire bombs" over Germany.[13] Advocates noted that the Germans had been prevented from using them against the Allies only by the direction of the prevailing winds.[14] Even if prevailing winds generally blew in a favorable direction for the Allies, it was by no means clear that balloons launched from Allied territories would reach their targets in Germany. Experiments with such weapons explored the accuracy of meteorological prediction, developed a balloon envelope sufficiently impermeable to hydrogen, and tested a mechanism for regulating the altitude of the balloon such that it took advantage of winds above about ten thousand feet. Test flights with the balloons involved the distribution of propaganda leaflets over enemy territory, but the Armistice was called before the Allied powers ever used them to drop bombs.

As Charles Ziegler notes, the "chief significance of the World War I balloon bomber lies not in its putative value as a weapon but in its status as the first practical unmanned vehicle for indiscriminate area bombing. Even if they were not operationalized, these experiments generated the conditions that gave rise to such 'terror' weapons considered as a type."[15] As such these experiments paralleled other attempts to operationalize the atmosphere, especially those with gas in World War I.[16] As Peter Sloterdijk has argued, the use of poison gas marked the beginning of a form of environmental warfare that targeted the elemental conditions of atmospheric forms of life. Gas is particularly insidious because it implicates in their own expiration the bodies of those it envelops. The balloon bomb is not so all-enveloping but is nevertheless implicated in the operationalizing of the atmosphere in order to target the material and affective infrastructures that sustain forms of life.

The promise of balloon-borne military operations was revived during World War II. In November 1944 the Japanese military began releasing balloons designed to drift across the Pacific. Equipped with incendiary and antipersonnel bombs, the hope was that the balloons might spark fires in the forests of the Northwest United States, diverting necessary and important resources from other theaters of war. Such weapons had been the focus of experiment by the Japanese since the 1930s; however, they were deployed only after the "Doolittle" bombing raid by the United States on Tokyo in April 1942. Even if the raid did not cause significant physical damage, it provided a boost to public morale in the United States. Balloon bombs offered the Japanese a possible counterweapon whose effects would be both destructive and psychological.

Filled with hydrogen, the balloons were fabricated from panels of mulberry paper (and in some cases silk) glued together by "Japanese school girls working in large theatres and sumo wrestling arenas in the Tokyo, Osaka, and Kyoto areas."[17] To make the five-thousand-mile journey across the Pacific, the balloons took advantage of the easterly jet stream blowing at altitudes above 30,000 feet. By releasing hydrogen or sand in order to compensate for the heating and cooling effect of sunlight and darkness, the balloons could maintain an altitude of between 30,000 and 38,000 feet.

An estimated nine thousand balloons were launched, only three hundred of which were confirmed to have made it across the Pacific, landing along a wide arc ranging from Arizona to Michigan to Alaska. As fire starters, the balloons were relatively unsuccessful: during winter and early spring the forests upon which they landed were too wet to be combustible. And of all the balloons launched, only one caused any fatalities, killing six people, including five children, on a church picnic in Oregon in early May 1945.[18] The story of the balloons' threat was not allowed to circulate widely in popular media. While some people, including the only survivor of the Oregon church picnic, were aware of the balloons, they never really generated a widespread atmosphere of fear among the civilian population. The Office of Censorship placed restrictions on media reporting of the balloons, while also informing service personnel of the gravity of the threat they posed.[19] Such restrictions were lifted with the deaths in Oregon, and even then only five weeks after the event.[20]

DISRUPTING INFRASTRUCTURES AND ATMOSPHERES OF SECURITY

In other theaters of wartime operation, the potential of weaponizing balloons through intentional release and drift was discovered only accidentally. Common sights in wartime Britain, barrage balloons were tethered around cities and strategic sites in the hope that they would deter and potentially down enemy aircraft flying lower than about five thousand feet.[21] During the night of September 17, 1940, a large number of these balloons were blown from their mooring sites on the southern coast of England by strong winds, drifting across the North Sea as far as Sweden. The mooring cables attached to the balloons snagged on power lines and communication infrastructures, causing power outages and disruption to transport networks. On December 18, newspapers carried reports of this disruption: "Many electric trains in Sweden were delayed as a result of damage by the trailing cables, and the wire dropping over the tracks made the use of steam engines dangerous. The central Swedish radio station's antenna at Motala was smashed, stopping the morning broadcast."[22] This event, coupled with complaints from power companies in Britain about the disruptive effects of wayward barrage balloons, convinced military authorities that they should experiment with their deliberate release as weapons of distraction and disruption. Further investigation showed that the German power distribution network was especially vulnerable to the powerful electrical arcs sparked by the trailing wires of balloons.[23]

Under the name Operation Outward almost 100,000 balloons were launched from sites near Felixstowe, Dover, and Great Yarmouth. Filled with hydrogen, the balloons were equipped with fuses that triggered a trailing line of piano wire to unwind, causing damage to anything below. The balloons were operated by crews drawn from Women's Royal Naval Service, each of which could launch up to ten per hour, depending upon the direction and speed of the wind at the estimated cruising altitude of the balloon. Under wartime reporting restrictions, the activities of the unit were not publicly known. In Germany, however, they had some success: immediately after the war the balloons were found to have inflicted considerable damage, causing a series of major disruptions to the power grid, and in one instance destroying a power station.[24] Evidence suggested a demonstrable impact that more than justified the small cost of the operation; indeed, in some cases the balloon release operations caused greater damage than more expensive bomber strikes.

Other balloon operations targeted the affective life of populations, as part of how, as Ben Anderson has shown, morale became the focus for forms of biopolitical and geopolitical intervention during World War II.[25] Created in 1938, the British M-Balloon unit operated under the control of Bomber Command with the purpose of distributing propaganda. Consisting of a rubber-coated cotton envelope with an open neck through which gas could escape when the balloon reached an optimum altitude, the balloons used by the M-Balloon unit had a fuse timed to release the leaflets, and were also designed to self-destruct at the end of their flight. At the beginning of the war operations were undertaken in France by teams that launched about twelve balloons per hour. With the fall of France the unit then moved to Cardington, England. The leaflets carried by and released from the balloons were produced by the Department of Publicity in Enemy Countries (DPEC) and later by the Political Warfare Executive, and contained various messages about the Nazi regime, in addition to texts of speeches by British politicians. The material distributed was initially straightforward, its origin and content obvious to whomever should find it. As Lee Richards notes, after late 1943 the content of this material became more subversive, often designed to appear as if it had originated from antiwar activists within German-occupied territory. Describing one such leaflet, the head of the Political Warfare Executive commented, "The whole leaflet, which is got up in the style of a soldier's news sheet, is intended to give a 1918 atmosphere to the Germans reading it and the general line is: 'The war is lost: better chuck out Hitler now rather than later.'"[26]

M-balloons were not precision weapons, and their final destination depended entirely on the speed and direction of the wind. Indeed, poor weather forecasting sometimes meant that leaflets were distributed up to 650 miles from their intended targets. But this was not always considered a problem. In reporting to the cabinet office, Sir Campbell Stuart, head of the DPEC, noted that the waywardness of these balloons could work both ways: "There is not, of course, the same control of direction; but even though numbers of the leaflets may be scattered over country areas, there is satisfaction in knowing that, in this event, they present a particularly embarrassing problem to the Gestapo. Moreover, there may be strokes of luck. On the 8th of November, for instance, a number of balloon-borne leaflets fell over the towns of Chemnitz and Freiberg at a time when the inhabitants were on their way to work."[27]

BALLOONCASTING

During the Cold War the balloon was also used as a device for operationalizing the meteorological atmosphere as an infrastructural medium for the distribution and dispersal of ideas and affects. From 1951 until the end of that decade, almost 350,000 balloons, containing 300 million leaflets, posters, and books, were released from sites in Germany and carried by the wind in the direction of the populations of the Soviet Bloc states of Czechoslovakia, Hungary, and Poland. Organized under the aegis of the Free Europe Committee (FEC) at the instigation of the Central Intelligence Agency, the balloon releases were coordinated closely with the broadcasting and programming of Radio Free Europe and Radio Liberty. These operations in Europe were linked, in turn, with a nationwide program of political mobilization known as the Crusade for Freedom intended to generate anticommunist feelings within the United States and to raise funds for operations within Europe. Backed first by President Truman, and then President Eisenhower, the balloon launches were framed publicly as a novel and necessary method for distributing messages of hope and freedom to the populations of the Soviet satellite states whose governments worked actively against the distribution of ideas by more conventional means. Commenting on the early launches of these balloons in 1951, the *New York Times* claimed that they represented the "newest way of getting an idea over the Iron Curtain."[28] One prominent syndicated columnist, Drew Pearson, who attended early balloon launches in Germany in 1951, wrote, "Near the Czechoslovak border, the current experiment in penetrating the iron curtain by balloons may be a great success or it may fail. It is too early yet to say. But the important thing is that it's an attempt by private individuals under the free-enterprise system to try out certain methods of psychological propaganda—or call it psychological warfare if you will."[29]

The balloon operations of the FEC became an important part of the atmospheric geopolitics of the Cold War and, albeit in a modest way, the technoscientific experiments that informed aspects of these geopolitics.[30] While radio was critical to these politics, anticommunist radio broadcasts had become intensely contested operations, not least because frequencies could be jammed.[31] Indeed, by as early as August 1948 the Special Procedures Group of the CIA had acquired a "printing press and a stockpile of meteorological balloons to carry and deliver propaganda leaflets."[32] At the same time, the members of Project Troy, convened secretly by the State Department, produced a report proposing that "an area of a million square miles could be saturated with a billion propaganda sheets in a single balloon operation costing a few million dollars . . . if the area of dispersal in such an operation were restricted to 30,000 square miles, which may be practicable, there would be a leaflet laid down, on the average, for each area of 30 by 30 feet . . . dispersion of balloons in flight and the dispersion of leaflets in falling from altitude both lend themselves to saturation operations."[33]

The first practical experiments by the FEC took place in August 1951 from a site in Germany near the Czech border. With "Svoboda," the Czech word for "freedom," printed on their side, the balloons carried leaflets proclaiming in capital letters that "A NEW HOPE IS BLOWING. A NEW HOPE IS STIRRING," and also reminded intended readers that "tyranny cannot control the winds."[34] Two types of balloon—plastic and rubber—were launched, both about four and a half feet in diameter. The former distributed its payload of leaflets by descending intact after a specified length of time; the latter, in contrast, burst at a measured altitude and scattered leaflets over a wide area. The sublimation of a specified amount of dry ice determined when the leaflets were released.

These operations also depended upon prevailing meteorological conditions and reliable knowledge of these conditions. During preparation for a balloon-borne leaflet-saturation campaign over Hungary, a "detailed study of the historical wind velocities and directions" was undertaken, which indicated that it would not be possible to count on "more than 25% operational weather."[35] On the ground, inflating and launching the balloons required care and attention, not least because they were filled with hydrogen. The kinds of activity at the site of the first launches in 1951 were described thus:

> It was extraordinary to see the operation work out. The green crews had to stuff exactly the right number of leaflets into each of the two different types of balloons, exactly the right amount of hydrogen had to be blown into each type, and the balloons had to be sealed. . . . The plastic balloons had to be measured by hanging weights on them until the balloon was in exact equilibrium. . . . After the first balloon was aloft the crews settled into a real production rhythm, and by the time an hour had passed things were really humming smoothly.[36]

Initial operations were undertaken independently of radio broadcasts but soon became aligned with focused radio-based political campaigns. Operation Prospero involved radio broadcasts coordinated with the release from a site near the Czechoslovak border of about 6,000 balloons during a four-day period in the summer of 1953. Operation Veto, undertaken in April 1954, involved the release of 63,786 rubber and 11,296 plastic balloons that distributed about 41 million pieces of printed material over an area of 36,000 square miles "containing an estimated population of 11,800,000."[37] Timed with radio broadcasts, these operations were intended to encourage "peaceful liberation" in targeted states by giving "shape to an unformed opposition."[38] In practical terms, this involved the distribution of leaflets through a balloon-borne saturation campaign coordinated with the schedule and content of twenty-four-hour radio broadcasts. As one report noted, "A typical example of the coordination of printed and spoken work operations after the initial phases of Veto was the publication by balloon of the Letter of the Winds on June 6. As the Letters of the Winds were released, the Voice of Free Czechoslovakia changed 12 of

FIGURE 8.3 Balloons being launched from a site in West Germany near the Czech border during the Radio Free Europe balloon operations in the 1950s. Source: Radio Free Europe/ Radio Liberty.

its programs to repeat the message contained in the letter to assure even more thorough saturation of the country."[39]

This enthusiasm for leaflets mirrored experiments with their use by other agencies during the early years of the Cold War.[40] Soldiers ordered to collect leaflets could hardly avoid reading them. And once found, the leaflets could be distributed by hand; it was assumed that the balloon-borne distribution of leaflets would then facilitate the distribution of a feeling of affiliation across different groups and individuals. As one evaluation report noted,

> The oppositionist will find a leaflet in the field: he will read it and re-member the message, keeping it to himself. Or he will tell others of the contents of the message. Or he will take one, or several, of the leaflets and distribute them among his friends and family, mail them to activists, stick them up on a wall or on a fence or three. . . . It is

entirely up to him how to act or what risks to take. But at the very moment he acts, he feels and knows himself to be a member of a larger group, a movement, and he gains self-confidence. He feels a sense of initiative and movement in his hands. In short, he is enlisted *actively* as a member of the Opposition.[41]

For those who designed and planned the operations, the point was to use the trajectories of the meteorological atmosphere to generate and modify political atmospheres in the countries targeted. Combining radio and leaflet operations enhanced "suspense-building," and the former could "amplify" the messages on the latter.[42] Underpinning these operations was the claim that "the two media combined create an atmosphere of urgency in which the effect is not doubled but multiplied."[43]

Such operations were not just designed to address an audience in Soviet satellite states. From the early 1950s the radio and balloon operations in Europe were linked explicitly with domestic morale and fund-raising activities in the United States as part of the decade-long Crusade for Freedom campaign. Backed publicly by many cultural and political personalities, the campaign was also shaped strongly by expertise from the newly formed Advertising Council. The emphasis here was not just on publicizing the operations in Europe but also on organizing local events to generate support across the United States for these operations. A report published in January 1950 suggested that "every known method of publicity should be utilized but the weight of the effort should be localized publicity."[44] The advertising was to be based upon principles of "emotion, reason, and action." With respect to the former, "personalized dramatic headlines will be written in terse simple words that speak directly to the average man."[45] Following this, the report continued, "once the emotions are aroused, the immediate danger is stated" in the hope that the "reader will be motivated to the action he is next exhorted to take."[46] Part of this involved an advertising and publicity campaign designed to encourage citizens to sign "freedom scrolls" and to donate money to support the activities of Radio Free Europe.

This campaign involved a range of modest activities in towns and cities across the United States at the center of which was the display of a traveling "Freedom Bell" carried on the back of a truck. These events included balloon releases. Indeed, in many cases, the fund- and morale-raising activities in the United States mirrored those taking place in Europe. In October 1950, for instance, a thousand balloons carrying freedom scrolls were released from the Empire State Building to demonstrate on a small scale the method used in Europe, in an act described by one newspaper as "filling the air with 'freedom' notes."[47] Many similar events took place, with prominent celebrities attending and endorsing the launches—Clark Gable, for one, launched balloons from Reno with instructions attached for how the finder might return them. Balloon races were held, with prizes for those who found the balloon that had traveled the farthest. Concerned citizens could even sponsor balloons. One advertisement stated, "Now for the first time your Crusade for Freedom has been able to intensify its tough idea war with Communist rulers with *written words—messages* carried in balloons blown by the 'winds of freedom' deep into the captive countries. This new and dramatic means of piercing the Curtain is a significant step."[48] To support this campaign, interested citizens could make monetary contributions, and could choose to have their name printed on the balloons.[49]

The level of donations provided one way of assessing the degree of domestic interest in and support for the FEC balloon operations in Europe. Assessing the impact of the operations in the countries targeted was a different matter. Interviews with individuals who managed to leave these countries were used "primarily as indicators of atmosphere and background."[50] For instance, an internal FEC report completed in February 1955 was based on the statements of forty refugees from thirty different locations within Czechoslovakia and indicated "widespread" knowledge of Operation Veto.[51] The report noted that leaflets were distributed by hand in Czechoslovakia. Another internal confidential report about the effectiveness of Operation Veto drew upon similar sources, one of which suggested that "Balloon leaflets (Operation Veto) very successful. People eagerly collect leaflets under pretext of picking mushrooms in woods. Women interested in articles on new American fashions; men in workers' standard of living in U.S."[52] And yet another quoted a source claiming that in Vycapy, a small village in Moravia, "most of the leaflets that came down . . . fell into the hands of the inhabitants who, in turn, gave them to people from other villages where few, or no, leaflets had fallen. . . . Soon a female teacher brought word into the village that many balloons had come down in the districts of Jihlava and Ostrava. Workers carried leaflets to work in their briefcases and distributed them among other workers from villages in which leaflets had not come down."[53]

The balloon operations undertaken in Cold War Europe were designed to be disruptive enough to unsettle carefully controlled media envelopes while remaining benign enough not to precipitate a conflagration.[54] This was a tricky balance to maintain, as other balloon operations undertaken during the 1950s had revealed. The outbreak of the Korean War in 1950 provided an important stimulus for research into the military value of the balloon. Central to this research was the RAND Corporation, established initially by the US Air Force in 1946. Led by William Kellogg and Stanley Greenfield, this research was inspired in part by knowledge of how Japanese balloons during the war had taken advantage of an "understanding of upper atmospheric meteorology."[55] It was also shaped by Kellogg's experience using high-altitude balloons to detect radioactive particles after atomic tests. What became Project Gopher was based around balloons fabricated from (the recently invented) polyethylene that could fly high enough to be invisible to the naked eye, radar sensors, and air defenses, while also carrying reasonably large payloads. Another report by Kellogg led to the establishment of a further balloon program, called Project Moby Dick, which involved experiments with high-altitude remote sensing and reconnaissance.

In 1951 the US Air Force assembled a group of experts to advise about methods of conducting reconnaissance over Soviet Bloc countries. Based in Boston, and organized under the aegis of MIT, the Beacon Hill study group investigated the possibilities of different technologies and also undertook site visits to "airbases, labs, and firms for briefings."[56] Their report, *Problems of Air Force Intelligence and Reconnaissance* (1952), was generally positive about the potential of balloon-based reconnaissance platforms. They considered many proposals, among which was an "invisible" dirigible, a "giant, almost flat-shaped airship with a blue-tinted, nonreflective coating [that] would cruise at an altitude of 90,000 feet along the borders of the Soviet Union at very slow speeds while using a large lens to photograph targets of interest."[57] At the same time, and alongside researchers at RAND, this report identified some of the political problems with balloon flights over "enemy" territory. Such problems were crystallized by another balloon project. In 1956 Strategic Air Command began a program of balloon-borne reconnaissance over Eastern Europe, the Soviet Union, and China under the heading of Project Genetrix.[58] Over a twelve-month period 516 balloons were launched, nominally with the aim of contributing to weather research as part of the International Geophysical Year. As an experiment in photoreconnaissance, however, the project was only partially successful. The winds blew the balloons in a more southerly direction than predicted, and some were either shot down or malfunctioned. As a result, photographic material was recovered from only thirty-four balloons. In addition, both the Soviets and their satellite states in Eastern Europe protested vehemently to the overflights. The adverse publicity convinced Eisenhower that such "weather balloons" were more trouble than they were worth. His disapproval raised fears among the relevant agencies that the propaganda balloon releases being carried out in Europe might also be cancelled, and that the ongoing development of the U-2 spy plane project might be halted.[59]

Eisenhower's caution was tempered somewhat by his enthusiasm for the technical and affective significance of other balloon projects, including Project Echo. As his voice transmission via the Echo balloon suggested, the significance of such objects in the sky, and the capacity to broadcast and transmit live voice signals via these objects, involved more than a technical relation with new forms of atmospheric media. It also had a geopolitical significance, much of which was about the affective atmospherics that accompanied the prospect of seeing something in the sky in the context of the Cold War.

HAILING

The political scope of the relation between envelopment and atmosphere is not exhausted by the vision of the future divined by Lana de Terzi. It is not captured fully by the watchful, panopti-scopic sphere, dramatized in the shape of *Rover* in *The Prisoner*. And it exceeds the promise of a net of aerostatically enabled sensors that provides the rationale for contemporary experiments in surveillance systems like the JLENS Orbit. As operations Veto and Prospero illustrate, the politics of atmospheric envelopes are far more wayward than this. This is because operations Veto and Prospero are assemblages defined by shifting relations between prevailing meteorological conditions, the properties and capacities of the devices and materials used, and the discursive and affective scripting of the printed matter carried. Their spatiotemporality is defined not in terms of the clearly demarcated volumetric spaces of territory but in terms of variable volumes composed from wind speed and direction. The operational logics of experiments with the atmosphere as an infrastructural medium exceed the premise of dirigibility upon which many envelopes of security are based. The logics of these experiments are defined instead by drift, distribution, and dispersal.

The politics of these operations can be understood in terms of the performance of a distinctive mode of atmospheric address. I mobilize "address" in two senses here. The origins of the word lie in the Latin *ad* (toward) and *directus* (direct). "Address" is therefore both a place and the principle of guiding something in the direction of something else. An atmospheric mode of address is not, however, about the point-to-point transmission of a voice message or speech to a specified individual at an identifiable location. It involves an act of release that enables the drift and dispersal of clouds of objects, ideas, and affects toward a vaguely specified destination. Moreover, an act of atmospheric address does not necessarily operate on the basis of the targeting of a population specified in advance, although it can, as the case of the operations above illustrates. Critically, it can also involve the release of things into the atmosphere that call forth an audience or constituency whose shape is not known or determined in advance.

One way to grasp an atmospheric mode of address might be through the concept of hail/hailing, and in two senses of that term. The first is meteorological. Understood thus, hail is a meteorological process of precipitating multiple entities through a form of heating and cooling and rising and falling in the atmosphere. Hail, in meteorological terms, is the precipitation of icy objects from the energetic turbulence of clouds. This meteorological sense can be stretched a little: it can provide a way to think about the distribution of other kinds of entities and how they become implicated in the precipitation of atmosphere affects. We might think, for instance, of a hail of shells, a hail of bullets, a hail of insults. The operational logic of experiments in release and drift can be grasped in terms of a modified form of hailing as a mode of atmospheric address: it involves the release of multiple objects, ideas, and affects that drift toward a target, albeit one that can be reached only if the meteorological conditions, particularly wind, are right. Together, these things travel not as an object or entity but as a kind of vague, diffuse, atmospheric event. In that sense, they have the quality of what Gilles Deleuze and Félix Guattari, following medieval scholars, call a "haecceity": by this they mean the intensive "thisness" of events.[60] We might also think of this as the voluminous intensity of an atmospheric thing—defined not simply by extension but also by the degree of intensive relations between elements in ongoing compositional formation.

Importantly, Deleuze and Guattari remind us that to name these events is not to turn them into objects or entities, or to reduce them to static representations; it remains possible to think of them as operational assemblages in formation, composed of different trajectories, speeds, and relations. As Deleuze and Guattari explain, "The proper name fundamentally designates something that is the order of the event, of becoming or of haecceity. It is the military men and meteorologists who hold the secret of proper names, when they give them to a strategic operation or a hurricane."[61] These are proper names such as Operation Outward, Project GHOST, Operation Veto, and Operation Prospero. All of these proper names indicate operations whose spatiotemporality is composed of relations between different elemental agencies and forces. While they are named, and while these names have performative force in the administrative and geopolitical assemblages that enunciate these names, the relations of which these operations are composed remain in flux, uncontained, and atmospheric, and in ways that are excessive of any entity. These operations, in other words, are atmospheric things: they have a loose, elemental consistency, but their voluminous spatiotemporality stretches out to form an envelope of variable shape and intensity.

Once released, these operations have the potential to address other forms of life through a second kind of hailing. To hail is also to call out, to solicit, to catch the attention of someone or something. It is a form of sounding, albeit one more directed, and one taking the form of an injunction of sorts. To be hailed is to be addressed or to be called to respond as the subject of action in some way. No. 2 does precisely this when he calls out "Halt, be still!" in that scene from the first episode of *The Prisoner*. But hailing is not only about addressing a subject or an entity as the bearer of action; it is also about addressing, or calling forth, something more vague, something that takes the form of an atmosphere from which affective variations can precipitate. This, indeed, is one of the aims of the various balloon experiments above; they are designed not so much to address individuals, or subjects, but to call forth affective variations, however minor, in atmospheres whose relative stability becomes the object-target of practices and techniques of security.

ACTS OF ATMOSPHERIC ADDRESS

Hailing therefore links the time, or *temps*, of precipitation with the sounding of an atmospheric mode of address, and it does so in the operation of a form of atmospheric politics that draws together the affective and meteorological senses of the atmospheric through practices and processes of envelopment. This politics operates interstitially to the operational logics of dirigibility and visibility performed by other, more recent airborne craft, including the drone. It has the potential to disrupt or at least complicate these logics because it involves the distribution of atmospheric things defined by the movement and force of variations in the meteorological atmosphere.

How then might small, modest acts of release operationalize the elemental variations of the atmosphere in ways that are not defined by the geostrategic visions of the examples above? To pose this question might seem rather quaint. Under the weight of atmospheres of security defined by dirigibility and persistent presence, the operationalizing of the meteorological atmosphere as a medium through which objects, ideas, and affects can be distributed may now seem somewhat anachronistic. However, minor acts of balloon release continue to be used as interventions designed to disrupt atmospheres of security at a number of sites. Borders are critical here. After all, the balloons launched as part of Operations Veto and Prospero were released near borders. Equally, the size and capacity of these balloons meant that it was extremely difficult for those charged with securing the integrity of those borders to prevent incursions into controlled airspace: it is difficult and expensive, although not impossible, to use a fighter plane to shoot down a balloon.[62]

Balloon release continues to be used at border sites to generate atmospheric things whose voluminous qualities disrupt, however modestly, the volumetric security envelopes of certain states. Consider, for instance, the well-publicized use of balloons in the demilitarized zone (DMZ) between North and South Korea. "I get choked up, every time, as I let go and watch it take off"—this is how pastor Eric Foley describes his experience of launching balloons from a point in South Korea near the DMZ. Foley and his wife, Dr. Hyun Sook, have been launching balloons since 2006 under the auspices of an organization called "Seoul USA" founded in 2002. Made from clear plastic, the balloons are about forty feet high and filled with hydrogen. They carry beneath them bibles and literature intended for the 100,000 underground Christians living in North Korea, and are equipped with timers that release the material at points identified by the launchers.[63] Escapees and defectors from North Korean have also used similar kinds of balloons to send messages across the border. Another Christian activist, Lee Min-Bok, has been sending Bibles and related material across the border since 2003. A former collaborator of Lee's but now something of a rival, Park Sang-Hak also releases balloons, but without the religious associations. Park is director of Fighters for a Free North Korea, an organization that since 2003 has sent approximately two million balloons across the border carrying ten million leaflets, in addition to USB drives and radios. In 2013 Park was a recipient of the Václav Havel Prize for Creative Dissent, awarded at a ceremony in Oslo.[64] Commenting on the prize, Park Sang-Hak remarked, "This prize is not for me alone; it is for the 25,000 North Korean refugees who are fighting against the dictatorship of North Korea; it is for the helium balloons that deliver our weapon of truth."[65]

Such launches are unpopular with the North Korean regime, which has at times threatened to target the launch sites.[66] The South Korean authorities also disapprove of these launches, and have taken active steps to prevent them—as have groups in the South who favor a conciliatory approach to relations with the North. However, after the sinking of the warship *Cheonan* in 2011, with the loss of forty-six sailors, the approach of the South Korean ministry became a lot more relaxed, and balloon releases across the DMZ are now tolerated more openly. At the DMZ, balloon release, nominally about sending messages of hope and freedom across the border to the north, becomes a political act that deliberately punctures the impermeability of the border by distributing objects, ideas, and affects. The windborne waywardness of the balloon enables a technique for unsettling the security atmospheres that envelop North Korea.

Elsewhere, near other borders, small acts of balloon release provide occasions for disrupting, albeit temporarily, envelopes of security and surveillance around territorial volumes that in some case are maintained, at least in part, by the kinds of systems with which this chapter began. Since the early 1980s, tethered aerostats have provided surveillance at various points on the border between the United States and Mexico, as part of the Tethered Aerostat Radar System that "guards" the southern border of the United States.[67] Each aerostat carries a radar with a range of up to two hundred miles in order to detect aircraft engaged in illegal activities. Smaller "tactical" aerostats are now being trialed by the Customs and Borders Protection agency following experience of their use by the US military in Afghanistan. These portable systems are designed to monitor activity on the ground and are equipped with infrared and thermal sensing devices. The Israeli Defense Force (IDF) uses similar systems at the borders between Israel and the occupied territories. Manufactured by RT Aerostats Systems, Skystar systems are used by the IDF Combat Intelligence Collection unit for surveillance and threat tracking.[68] They formed a crucial part of the surveillance aspect of the Gaza War of 2014, in Israel officially named Operation Protective Edge.[69]

The presence of these systems reminds us that borders are not so much narrow lines of separation but more extensive envelopes of surveillance and security. They also raise questions about the kinds of interventions that might disturb these envelopes. In both cases, albeit in different ways, small acts of release can provide some possibilities. For instance, the Palestinian National Authority released 21,915 black balloons in May 2008 as part of a series of events to mark the sixtieth anniversary of Israel's declaration of statehood. The number was chosen to symbolize the number of days since the occupation of Palestinian territories began.[70] Smaller-scale balloon releases are often used to mark events and occasions at walls and fences that fracture Israeli- and Palestinian-controlled territories. These are minor events of atmospheric address that generate modest atmospheres of collective volume, and often in ways that are linked with similar events taking place elsewhere.[71] This act has also of course become linked with the artist Banksy's (2005) graffiti work—"The girl with balloons"—on the West Bank Wall. Such acts are also depicted cinematically in Elia Suleiman's film *Divine Intervention* (2002), in which a character played by the director releases an orange-pink balloon—with the smiling face of Yasser Arafat on it—to drift across a border checkpoint in a way that perplexes and confuses the soldiers below. As Patricia Pisters writes, in this scene, "a very simple balloon (its journey digitally enhanced by CGI to travel in perfect accordance with the filmmaker's wishes) can transform into a powerful image of resistance: and in the same moment, we truly perceive the fragility of the kind of resistance art can provide."[72]

Elsewhere, other artistic deployments of this device reveal subtly different ways of thinking about the politics of balloon release. Similar events have also taken place at other borders. In 2000, as part of a work called *The Cloud*, Chilean-born artist and architect Alfredo Jaar released a large number of white helium-filled balloons from a site near the US-Mexico border (figure 8.4). This work, which also involved the performance of a concert at the site, was intended to memorialize migrants who had died in the act of trying to cross this border.[73] For a while the balloons, clustered together, were tethered near the border, looking, however unintentionally, like a surveillance aerostat. Once released, the balloons drifted off, scattering.[74] However, as Nicole Sheren notes, they "floated not toward the United States, the direction of the prevailing winds, but instead on a gust toward Mexico. As is evident from his project proposal, Jaar had intended for the cloud to float toward the United States, in a final successful border crossing for the dead. Instead the freed balloons moved south, lending a different layer of symbolism to the work."[75] In effect, the balloons, and their wayward movement, generated what Sheren calls an

FIGURE 8.4 Alfredo Jaar, *The Cloud*, 2000. Public intervention, Valle del Matador, Tijuana-San Diego, US-Mexico border. Courtesy of the artist.

"impermanent memorial" in the air to those who had died trying to cross the border.[76] As they did so, they also generated a cloud defined in part by the intensive spatiotemporality of the wind.

It would be easy, of course, to exaggerate the political significance of such works. They are minor, in many senses of that term, operating transversally to established forms, spaces, and surfaces of politics and to the territorially defined atmospheres of security on which these politics are premised. They do not so much target populations as much as they hold out the promise of calling forth minor publics through small acts of address. Nevertheless, such acts are worth thinking about as incursions into regulated airspace and security atmospheres in ways that are relatively cheap yet difficult to prevent. However modest, these works operationalize the lower reaches of the meteorological atmosphere in ways that both recall and go beyond the vision of Lana de Terzi and other, similar attempts to deploy the balloon as a weapon. Here, instead, relatively benign acts generate localized atmospheres of attention with the potential to catalyze unpredictable affects. These are ways of moving with the directions and vectors of the elements whose apparent simplicity poses a challenge to highly complex systems of surveillance. The logic of these acts catalyses drift as a mode of atmospheric address: what travels, in the process, are atmospheric things composed of multiple objects, ideas, and affects that move, windblown, with the potential to hail a diffuse, unspecified addressee.

9   ELEMENTS

Sometimes you spend so long thinking about the same thing that it begins to disappear from view, in the same way that a word repeated often enough almost loses its meaning. Sometimes you spend so long trying to hold on to something without holding it down that it becomes too heavy to bear. So it is with the atmospheric things to which this book has returned again and again: at times they seem so vague and intangible so as to evaporate, becoming inconsequential, while at others they almost collapse under their own weight and shape and gravity. Perhaps residing in the tensions between these possibilities is an unavoidable consequence of thinking atmospherically. Perhaps atmosphere only offers a way of grasping the dispersed and diffuse materiality of the milieus in which bodies feel immersed because it is often only possible to grasp this materiality via something which does not itself have the quality of an atmosphere.

There is a parallel here between the atmospheric and the elemental. The former, indeed, marks an important issue through which a renewed interest in the elemental and the elements has resurfaced across a range of disciplines and genres of thought.[1] Like atmosphere, the elemental, on one hand, is an expansive, turbulent field of often barely perceptible processes. It is an environmental milieu in which forms of life or entities are conditioned and immersed, one whose variations, taking place at myriad scales and degrees of intensity, can sometimes be felt within and across bodies. Disclosing this sense of the elemental is central to the allure of envelopment; envelopment, after all, is a process through which a relation of actionable, palpable difference takes shape within an elemental milieu. In the shape of the balloon as a technical device, envelopment facilitates the sounding of the atmosphere in which it and other bodies are immersed. In turn, this atmosphere becomes the object of a range of calculative and anticipatory practices across many domains of expertise. At the same time the process of envelopment can and has been operationalized in more obviously aesthetic ways in order to generate elemental conditions from which diverse atmospheres of experience precipitate, glimpsed in works as diverse as Christo's *Big Air Package*, Warhol's *Silver Clouds*, Forsythe's *Scattered Crowd*, and Saraceno's *In Orbit*.

Even as they draw attention to the elemental as environmental condition and milieu, experiments with the balloon can also foreground other senses of this term. These are also experiments with the fabrics from which balloon envelopes are fashioned and the gases with which they are filled. They are experiments with materials such as Mylar and polyethylene, products of the revolution in industrial chemicals throughout the twentieth century. They are experiments with gases such as hydrogen and helium, both essential at different points to the properties and capacities of the balloon as a technical, aesthetic, and geopolitical device. Such gases have also figured here in other, less obvious ways. Helium, for instance, was central to the sensing devices that facilitated both Project Echo and the experiments by Arno Penzias and Robert Wilson: the receiver on the Holmdel antenna was cooled by liquid helium, sourced, in all likelihood, from an underground storage facility near Amarillo, Texas. As the narration to the film *The Big Bounce* put it, "This is no ordinary amplifying device. It operates in an ultracold vat of liquid helium at minus 457 degrees Fahrenheit. In this supercold world, the maser's ruby crystal helps keep the telephone conversation clear and easy to hear."[2] These experiments are reminders that the elemental also refers to particular physicochemical configurations of matter and energy arranged diagrammatically in the form of the periodic table. To acknowledge this is not to reduce the elemental to the terms of a narrowly defined scientific gaze; nor is it to invoke a mode of brute, atomistic materialism. It is to recognize that the properties and capacities of particular elements, variously engineered and elusive, can work to shape, constrain, and enable different forms of life to different ends, both malign and benign.[3]

There is a third sense of the elemental, however, one also disclosed through speculative experiments with the balloon as a device for doing atmospheric things, and one especially germane to an account of the allure of envelopment: this is the elemental as a metaphysical proposition about the ontological composition of reality. Such propositions inform classical conceptions of the world. In accounts inspired by Empedocles, for instance, the world is composed of states of matter such as earth/solid, water/liquid, air/gas, and fire/plasma.[4] The tensions between more recent versions of a philosophical elementalism form the backdrop for some of the key claims made in *Atmospheric Things*. Focusing on envelopment provides a way of thinking and moving between, on the one hand, speculative realisms in which the entity—and sometimes the object—is the elemental point of departure for accounts of reality, and, on the other, forms of materialism in which process is the elemental condition for any realistic account of how things come to be the way they are. This means borrowing from an atmospheric materialism an understanding of the elemental as a condition excessive of an entity whose variations can be sensed in bodies of different kinds. At the same time, it means borrowing from speculative realists such as Graham Harman the claim that the *allure* of the elemental resides in its withdrawal from apprehension. From this claim emerges the wider proposition that the allure of envelopment resides in how this process generates entities that draw attention to and disclose the force of an atmospheric elementality excessive of these entities.

I hold on to the balloon here one more time as a speculative device for thinking about the allure of elemental envelopment. I follow it via experiments into the stratosphere while also thinking about its possibilities for crafting new forms of open-source atmospheric awareness on the ground. These experiments are not yet fully formed. They are pilot projects of a sort, and not only in a technical or scientific sense. They provide important opportunities for thinking about how processes and experiences of envelopment might become implicated in different apprehensions of the ethics and politics of the elemental in which the latter is simultaneously milieu, entity, and ontological proposition. At stake in these experiments, however modest, is the question of how to develop and distribute capacities to understand and move with variations in the elements as part of the elaboration of new forms of atmospheric life.

ELEMENTAL INFRASTRUCTURES AND ATMOSPHERIC MEDIA

Viewed from a point external to this planet, the earth's atmosphere appears as a discrete thing. It is a thin, limited envelope of gases surrounding the earth, akin to, but with properties different from, the atmospheres that surround the other planets and some of the moons in this solar system. And yet when we zoom in we see that this entity is itself composed not of other entities, but of processes including, but not limited to, radiation, absorption, convection, precipitation, condensation, and evaporation. The effect of these processes is to give the atmosphere a layered character. The layer in which we live, in which we are exposed to the meteorological elements, where we inhabit what Tim Ingold calls "weather worlds," is the troposphere.[5] Above a boundary zone called the tropopause is the stratosphere, that layer of the earth's atmosphere which begins at about eight to fifteen kilometers (depending on latitude) and extends up to about fifty kilometers.[6] Unlike the troposphere below, the stratosphere is characterized by a gradual increase in temperature with altitude, caused by the absorption of ultraviolet radiation. This inversion is crucial because it means that unlike the turbulent vertical movement that characterizes the troposphere, the movement of air in the stratosphere is largely horizontal, composed of trajectories moving at different speeds and altitudes. As a result, the stratosphere has little of what we might call and feel as weather.

Sounding, as a process of remote sensing, was and continues to be central to the disclosure of the properties and dynamics of the stratosphere. At the turn of the twentieth century, about 140 sounding balloons launched by the French meteorologist Léon Teisserenc de Bort made it into what he identified as the "isothermal layer" above which temperature no longer decreased with altitude.[7] Coining the term stratosphere, de Bort also realized that "we cannot theorize on atmospheric phenomenon as if they were continuous in time and space."[8] During the twentieth century the stratosphere continued to be the focus of a range of experimental balloon operations. Many of these were modest in scale, but some, including Project GHOST and Project Skyhook, involved more complex and distributed assemblages of entities, practices, and devices. Others involved the ascent by humans into the stratosphere in capsules and pressure suits of different degrees of safety and sophistication. These latter flights went some way to allowing the stratosphere to enter into the public imagination, with ascents by Auguste Piccard, Jean Piccard, and Jeannette Piccard in the 1930s, and by Malcolm D. Ross in the 1950s, contributing to a growing awareness, particularly in the United States, of its rarified, almost otherworldly, existence.[9] At the same time, the stratosphere became the focus of geopolitical competition through the launch by both the Soviet Union (in the 1930s) and the United States (during the Cold War period) of high-altitude balloons. More recently again, the stratosphere has become a matter of public concern in various ways: through controversies about the depletion of ozone; through proposals for it to become the target and medium of various forms of geoengineering; and, rather more trivially, by the heavily sponsored ascent of Felix Baumgartner via balloon into the stratosphere from where he undertook a skydive in 2012.

For well over a century, then, the stratosphere has been an experimental zone in the elemental milieu of the atmosphere. In mid-2013, these experiments once again drew a degree of public attention, when Google announced that it had been launching balloons into the stratosphere over New Zealand. Under the name of Project Loon, the aim of these launches was to test the feasibility of using balloons as High-Altitude Platforms (HAPs) for the distribution of wireless Internet coverage to parts of the world that lacked conventional terrestrial infrastructures.[10] After further tests, Google announced agreements in 2016 with telecom agencies in Indonesia and Sri Lanka.[11] At the time of writing, in mid-2017, Loon remains an ongoing experiment. Indeed, because they are equipped with ADS-B receivers, I can use a simple flight tracking software like *flightradar24.com* to find some of these balloons. As I sit here I can see a cluster of these balloons off the west coast of Peru: HBAL945 is at 63,700ft, HBAL156 at 63,100ft, HBAL189 at 63,000, HBALO15 at 58,900ft, HBAL187 at 62,900ft and HBAL247 at 62,000ft.[12] The availability of such tracking points to the extension of networks of visibility into the atmosphere: but it also reminds us that there might be more going on above us than we sometimes realize.

On the face of it, this experiment seems relatively benign: after all, it is surely a good thing to extend Internet coverage to places where it is limited or nonexistent. It is also difficult to critique this experiment from the perspective of parts of the world where such coverage is taken for granted and is provided by for-profit operators. Equally, as a relatively minor

FIGURE 9.1 A photo from the Google Loon launch event, June 2013.
Photographer unknown. Wikimedia Commons.

element of Google's scope and reach, the significance of Loon should
not be overstated. This is an experiment that could fail, and that might
amount to little more than a minor footnoted "moonshot" in the history of a corporation in which failure is of course often presented as a
virtue.[13] Nevertheless, when viewed through the lens of critical scholarship about security, Loon seems aligned uncomfortably closely with the
imperatives of the aeromobile gaze through which regimes of security
are underpinned.[14] It also raises questions about the elaboration of new
volumetric spacetimes, and of the modes of atmospheric governance and
ownership these volumes demand.[15] Loon raises these questions because
it is less a collection of individual entities than a moving assemblage of
devices, forces, and relations taking shape on the ground and in the air.
In moving through the stratosphere this kind of assemblage draws together the problems of territorial jurisdiction, regulation, and the legal
status of partially dirigible entities in volumes of air that are not easily
partitioned.[16] It is therefore no coincidence that the location chosen for

the first major trial of the Loon network—Christchurch, New Zealand—was also the same site from which, during the 1960s, the GHOST balloons operated by the National Center for Atmospheric Research were also launched. In both cases, this location reduced the likelihood of balloons overflying countries such as China and Russia, while also making air traffic control less complicated. But even the legal status of the Loon balloons is uncertain. Under Federal Aviation Authority and International Civil Aviation Organization guidelines, the rules for launching and operating these devices vary depending upon whether the balloons are classified as "unmanned" aircraft, drones, aerial vehicles, or aircraft systems.[17]

Speculating about Loon also leads us in other directions, however. This is a project that foregrounds how envelopment facilitates new ways of organizing the technical platforms upon which contemporary forms of media depend through reconfiguring the relation between infrastructure and the elements. To some extent Loon can be grasped, in the words of the project's technical lead, Baris Erkmen, as an attempt to "take a terrestrial infrastructure and sort of lift it up to the stratosphere."[18] While correct on one level, this depiction of Loon as the reproduction aloft of existing infrastructural arrangements underestimates its novelty. Certainly, if we understand infrastructure as a form of ontological experiment, then Google is involved in more than the replication of terrestrial infrastructure in the stratosphere.[19] It is better grasped as an experiment in making the stratosphere itself into a kind of elemental infrastructure. This is because it is premised upon the possibility of making stratospheric winds into streams of movement for the distribution and circulation of networks of devices that extend the connectivity of contemporary media atmospheres.[20]

Critical to this infrastructural ambition is the use of stratospheric wind trajectories to achieve a degree of technical dirigibility. Ground operators at Google headquarters in Mountain View, California, modify the altitude of the balloons, thereby allowing them to take advantage of winds moving at different speeds and in different directions. This is possible because the balloons in the network are in fact composed of two envelopes: an outer balloon contains helium to generate the lift to take it into the stratosphere, while an inner balloon (or ballonet) can be filled or emptied of air via a system of valves and fans that mean the balloon can ascend and descend.[21] The network draws live data sets from the National Oceanic and Atmospheric Administration that predict wind trajectories up to fifteen days ahead, albeit with diminishing degrees of accuracy. The analysis of this data by algorithms designed by Google engineers allows Loon operators to estimate where and when balloons in the network might find the right winds to take them to an intended destination; according to some reports it is now possible, under certain circumstances, for Loon balloons to circumnavigate the globe to within five hundred meters of a specified target.[22] At the same time, the technical feasibility of the Loon project depends upon the capacity of each balloon to remain part of a distributed, mesh-like communication network.[23] As one project member puts it, each of these platforms is in motion, "free-floating . . . swaying and bobbing," to the extent that the relations between balloons in the network can be understood as a "dance if you will between the two balloons where the lasers are trying to keep locked onto one another as they drift in the sky.'"[24]

Loon is an infrastructural experiment in which the elemental is not merely the material substrate on or through which infrastructure operates; instead, the elemental agencies of heat, light, and wind at play within the stratosphere generate the infrastructural conditions for the operation and arrangement of a complex assemblage. As Rich DeVaul, one of the engineers on the project, puts it, "We're using the sunlight, we're using the wind, we're using all of these things to build this network in the sky."[25] Or as another project member, Astro Teller, explains, "We can sail with the wind, and shape the waves and patterns of these balloons, so that when one balloon leaves, another balloon is set to take its place."[26] The promise here is both infrastructural and choreographic: what is choreographed is the complex technical arrangement between the force of the elemental and the capacity of envelopes to respond to and move with this force. This kind of choreography draws in the increasingly complex "environmental fringe" of media, which, in the words of Mark Hansen, constitutes a "crucial element in the rhythmic compositional dance that is our multi-scaled, disparate agency in the world."[27]

Recently Loon's choreographic capacities have become more complex. In earlier versions, Loon balloons circumnavigated the globe. In traveling, the Loon balloons were, in effect, performing a particular mode of standby in motion, with operators waiting for them to arrive at a point where they could become active participants in a distributed network of relatively limited extent. The technical feasibility of the project depended on a network of balloons sufficiently dense so that as soon as one balloon left another arrived to take its place. However, in early 2017 the project team announced it had found a way of changing how the balloons operate. Rather than drifting around the globe, the balloons could now be made to "loiter" in the vicinity of an area where they were needed. Writing in February of 2017, Teller, "captain of Moonshots X," noted,

> By early 2016, the team was seeing a few balloons behave in a slightly weird way: lingering in an area rather than sailing away. In the weirdness, they saw opportunity. They asked themselves the once-impossible question: could our algorithms help the balloons to stay much closer to the location they were already in? In mid-2016, we started sending balloons from our launch site in Puerto Rico to hang out in Peruvian airspace—and they did, some for as long as three months. We repeated the experiments, and saw the same results: we had figured out how to cluster balloons in teams, dancing in small loops on the stratospheric winds, over a particular region.[28]

In its more recent iteration Loon operationalizes an enhanced technical capacity to remain relatively still within an elemental milieu using the algorithmic coordination of the altitude of balloons in relation to the trajectories of winds within that milieu. Instead of hand-coded algorithms that respond to predetermined variables such as wind, altitude, or location, Gaussian machine-based learning is increasingly used to extend the operational capacities of Loon through forms of stochastic optimal control.[29] In the process, the data generated on previous Loon flights provides an archive for establishing a predictive horizon for the future movement of the balloons.

The technical capacities of Loon complicate any straightforward opposition between stillness and motion, and in ways that take this relation beyond phenomenological experience. And this is because Loon's operational present goes beyond the domain of this experience. It is less about how the feeling of stillness in motion is sensed in human bodies than it is about the distributed technical capacity to sense the conditions that make possible relative slowness in relation to the speeds and trajectories of an elemental milieu. Loon is about the technical capacity to condense and choreograph a loosely defined atmospheric thing from the midst of elemental movements. Loon is about harnessing value from something that, "has no summary form: it is a matter of pressures, flows, and frictions," a force that moves "around and through efforts by authoritative discourse and knowledge to orchestrate it as a simple 'object of analysis.'"[30]

Loon might be seen to presage the algorithmicization of any environmental variation, or foreshadow the rise of robots that generate new technical capacities for adequate directionality within conditions of partial predictability. At the very least, Loon foregrounds how practices of elemental envelopment might become infrastructural to forms of atmospheric media that are now part of the background of life, operating below thresholds of human sensing while also possessing an enhanced capacity to generate affective foregrounds as spacetimes of variable intensity and intensity.[31] If we follow thinkers such as Alexander Galloway and Eugene Thacker, notably, this might mean paying more attention to the elemental aspect of media that resides in the "ambient" or "environmental" aspect of their networks—that is, "all the things that we as individuated human subjects or groups do not directly control or manipulate."[32] These media networks have a spatiotemporality defined by condensation, precipitation, and dispersal, and need to be grasped through a "climatology of thought" that almost imitates the conditions with which it engages. In these terms media are elemental insofar as they operate in a way analogous to the atmosphere; they operate as a diffuse field of relations that can be configured technically in ways that precipitate localized clouds of affective intensity.[33] They have an ontology that is cloudlike insofar as it condenses and dissipates according to prevailing conditions and circumstances.

The ontology of such atmospheric media is elemental in a way that is more than metaphorical because they incorporate meteorological variations in their operational capacities. Loon is a reminder—if we needed reminding—of the continued importance of a more archaic and elemental sense of media as medium: here media are not simply channels of communication but environments, conditions, fields, and milieus, to whose variations bodies respond and to which they can contribute in a process of *mediation*.[34] As philosopher Levi Bryant writes, the concept of media "refers not to something that relates to the five senses—though that too—but rather to any relation between machines in which one machine *mediates* the structural openness, movement, or becoming of another machine."[35] In these terms Loon is an elemental infrastructure open to, and mediated by, variations in the stratosphere, variations that can themselves be understood as infrastructural. Operating as an experimental, elemental, infrastructural assemblage composed of wind, code, and craft, Loon reveals how the milieu in which forms of atmospheric media take shape is, following Michel Serres, a "mingled given" composed of circumstantial combinations of hardware and software.[36] And, as an elemental infrastructure premised upon the possibility of achieving a form of partial dirigibility, Loon foregrounds the importance of paying as much attention to how atmospheric "media steer and stay afloat" as to "the cargo they bear."[37]

To think with Loon in this way is to risk becoming aligned too closely with the speculative promise of a project in which technical experiment is presented as a kind of joyous, affirmative process of generating capital value. Perhaps this is a risk worth taking, however: these kinds of experiments provide occasions for speculating about the emergence of new kinds of conditions and objects for thinking.[38] The shape of these experiments encourages us to imagine a wider elemental commons in which a diverse range of operations and travels might take place. This commons does not yet, and perhaps cannot ever exist: it may only ever be a speculative space-time defined by the promise that different senses of the elemental—the ontological, the environmental, and physicochemical—are worked together as part of new arrangements of atmospheric life.[39] This might be a commons of different senses of the elemental that are related but not necessarily reconcilable.

The commons, of course, is a concept that already carries a great deal of weight. It invokes a space of which diverse constituencies partake without necessarily doing so according to a shared set of rules.[40] Certainly, to invoke a notion of the elemental commons is not to conjure up an aerial version of a space of shared togetherness, or to dream of a romantic atmospheric pastoral of which multiple species partake. And obviously, the fact that the elemental commons is as tragic as it is hopeful can be grasped in the way that different elemental milieus, including water, air, and rock, are zones for the disposal of toxins, pollutants, and detritus of myriad kinds. For generations, of course, the elemental commons facilitated a kind of collective forgetting of material excess. Now, however, there is plenty of evidence of the memory, and storage capacity, of this taken for granted background: evidence of how the accumulation, distribution, and aggregation of varying quantities of materials generates shifts in prevailing patterns, probabilities, and trajectories that are only beginning to be grasped.

All the more important, therefore, to speculate with the elemental commons as a means to generate different kinds of value in the atmosphere. Loon points to possibilities in this respect, but also reminds us of certain limits. In its operation Loon discloses how crucial it is to hold open the elemental commons as a medium of open-source experiment rather than a kind of privileged domain for generating capital-value. The dream of Project Loon is one in which atmospheric variations are transformed into sources of infrastructural capital, as part of a wider process of environmental value generation. If let run, this dream has a certain inflationary tendency. It points to an amplified form of what James Ash calls atmospheric "envelope power" premised on the capacity to activate elemental agencies at a range of scales, that stretches from screen to stratosphere, and whose medium is as much sky as silicon.[41] A specter of this form of power is conjured in that scene from *The Prisoner*, glimpsed in the previous chapter, in which envelopment is a suffocating condition of traceability, enclosure and death. Perhaps it is too easy to become too paranoid, however. The real issue lurking here is perhaps less dramatic: it is about how the technical capacity to experiment with infrastructures for elemental envelopment remains as non-proprietary as possible. Patenting devices and technologies diminishes this capacity. As Nick Shapiro has argued, the number of patents filed by Google in relation to Project Loon draws attention to how the accessibility and availability of technologies of envelopment for open-source experiments with being and becoming airborne might be reduced.[42]

This should not be mistaken, however, for a claim that the stratosphere, or the atmosphere more generally, should remain unconditionally open. It is a call instead to speculate on the infrastructural implications of operationalizing variations in the elemental milieu we call the atmosphere, and in its different layers, including the stratosphere. Such speculation might also gain support from other ideas of infrastructure. As Lauren Berlant has written, infrastructure is important for many reasons, but foremost among them is the fact that it is "that which binds us to the world in movement and keeps the world practically bound to itself."[43] Because of this, she continues, "one task for makers of critical social form is to offer not just judgment about positions and practices in the world, but terms of transition that alter the harder and softer, tighter and looser infrastructures of sociality itself."[44] Berlant points us to the importance of experiments that revise and remake the ontological life and promise of infrastructure; that is, experiments that generate conditions in which these infrastructures can become sources of new forms of value. Her concerns are very different from those animating this book. However, Berlant reminds us that the promise of speculating about infrastructures for an elemental commons is not so much that they help us construct a space of connectivity or commonality, but that it may allow us to fabricate novel infrastructural arrangements, albeit modest, for generating what AbdouMaliq Simone calls new "movements in perception."[45] And these movements, in turn, might help us sense and feel the elemental conditions of atmospheres anew.

Part of speculating about an elemental commons involves developing an awareness of how this commons might become articulated through experiments with the technical capacities of a range of devices (including the balloon). Some of these experiments, like Loon, are remarkably complex. Others are much more modest, involving small, localized acts of release. If winds can become infrastructural, then any act of releasing a balloon into the atmosphere is a minor experiment within an elemental commons. The possible entanglements of this act are already increasingly public as matters of concern. Indeed, while it remains alluring for all kinds of affective reasons, balloon release is increasingly contested, policed, and in many instances prohibited. A simple if obvious connection is now being made between the release of a balloon into the atmosphere and the fact that at some point, somewhere, this balloon will show up, surface, or materialize in a different form. Once released, a balloon will ascend and, depending upon the material from which it is fabricated, will become more brittle, perhaps eventually shattering into many pieces. Some of these pieces will fall back to Earth and become entangled, sometimes lethally, in other forms of life. Increasingly, then, and shaped by a range of public campaigns, balloon release in some jurisdictions at least, is considered akin to dropping litter on the ground.[46]

It is not enough therefore to affirm the windblown, windborne object as a figure through which to pursue a speculative project; it is also always necessary to think about the conditions under which the release of this object can be affirmed. This is not just a technical question, although the technical is a critical dimension of any answer. It is also political because it is about how the object, its associations, and the milieu in which it travels are governed. It is also ethical because it is about the forms of life that such acts might help cultivate, the forms of value they may generate, and their capacity to prompt new terms of engagement with the elemental. And it is also aesthetic because it involves thinking about how far and via what means the elemental conditions of atmospheres can be sensed by different bodies.

How then to draw together the ethics, politics, and aesthetics of the promise of release and its diverse elemental implications? Practices such as art and architecture offer possibilities in this regard. There are many examples of previous and ongoing experiments in art and architecture intended to make the elemental qualities of atmospheres explicit or to imagine forms of living in the air.[47] Sometimes these experiments engage directly with the promise of elemental infrastructures because, as architect Rachel Armstrong suggests, these infrastructures "may serve . . . to open up the possibility of designing in new and surprising places."[48] In the process, as Etienne Turpin writes, artistic experiments can instigate "renewed attention to infrastructures as the basis for the emergence and articulation of polities."[49]

Where then might we find sources for renewing the promise of release via experiments that use elemental infrastructures in order to generate new kinds of atmospheric thing? One such source might be the work of Tomás Saraceno, whose installation *In Orbit* figured in chapter 7. Saraceno's work has become the focus of a range of important engagements with issues of atmosphere and the elemental by scholars including, perhaps most notably, Sasha Engelmann, Bruno Latour, Andreas Philippopoulos-Mihalopoulos, Nick Shapiro, Bronislaw Szerszynski, and Etienne Turpin. Saraceno's work is multiple and manifold, but as the scholars above have observed, central to it is the relationship between envelopment and the elemental. In a range of installations and works envelopment is a process through which the elemental qualities of atmospheres become sense-able and palpable through forms that hold in tension insides and outsides.[50] Saraceno makes particular use of a distinctive kind of balloon—the solar balloon—which forms the basis for many of what he calls aerosolar sculptures. The solar balloon is distinctive because it does not require any heat from terrestrial forms of combustion, nor does it use hydrogen or helium to generate lift. Instead, the air within it is warmed by ultraviolet radiation from the sun during the day and infrared radiation from the earth at night. As Sasha Engelmann and I have noted elsewhere, in Saraceno's work the solar balloon becomes a craft for experimenting with envelopment as a technical and aesthetic process for sensing and responding to the elemental force of solar and terrestrial energies.[51]

The launch of a solar balloon/sculpture is not an end in and of itself. This act becomes an alluring event around which forms of collective experiment for an elemental commons might be re-imagined. Informed by earlier speculative and artistic experiments with the possibilities of devising forms of life in the air, in Saraceno's work the process of envelopment and the promise of release become components of a more expansive atmospheric imaginary, articulated through concepts like *Cloud Cities* and, more recently, *Aerocene*.[52] The latter, a deliberate variation on the concept of the Anthropocene, is a speculative experiment with crafting forms of life in the air that rely as much as possible on the elemental energies of the sun rather than carbon. Every launch of a solar balloon can be grasped as a pilot project for a wider elaboration of an elemental commons. Every launch performs a circumstance-specific piloting that promises to pull together bodies, devices, and concepts into new collective infrastructures in the air. The aim here is not so to represent the real elemental conditions of the present; rather, these experiments perform a "piloting role" in that they "construct . . . a real that is yet to come, a new type of reality."[53] The launch and release of a solar balloon becomes a speculative aesthetic experiment that invites coparticipation with new forms of elemental envelopment as the generative preconditions for inventing new spaces of atmospheric life. At a time when the figure of the pilot, particularly via the drone, has become a cipher for contemporary forms of aerial violence and surveillance, these experiments carefully affirm piloting as a responsive craft of becoming attuned to variations in the elemental conditions of the atmosphere. The emphasis here is not simply on technical achievement or ascension but on how the process of flying—or more accurately floating—with a solar balloon could extend our sensory capacities through a kind of speculative ecology of drifting. Here, Saraceno echoes aspects of earlier aerostatic excursions through a form of experimentalism that recalls distinctive experiences of being aloft: experiences of stillness and motion, of sounding, of becoming voluminous.[54]

These launches remain pilot experiments because only a limited number of free flights of Saraceno's solar works have taken place. In addition to technical limitations, the release of such craft into the air is subject to all kinds of legal and political restrictions. Yet, even then, such craft retain the potential to effect what, following Félix Guattari, we might call the resingularization of the act of release of something into the air.[55] What this means is that as part of the wider speculative lure of concepts like *Aerocene*, the act of release, which has been rehearsed to the point that it no longer seems to qualify as an event, can now generate new forms of thinking, feeling, and association.[56] This can begin to happen on the ground before anything ever takes to the air, something exemplified by one of the projects in which Saraceno is a collaborator. *Museo Aero Solar* is a participatory entity-event consisting of a solar balloon envelope fabricated from plastic bags (figure 9.2). In existence for about fifteen years, it has been installed at a range of sites around the globe, at which individuals are invited to contribute plastic bags for the envelope and to participate in the process of its ongoing fabrication. This work of fabrication—involving gathering, cutting, sealing, and shaping—rehearses while also reworking the activities through which many balloons have been fashioned, while not being circumscribed by the divisions of labor that characterized the fabrication of many earlier envelopes. In a kind of collective, collaborative fashioning, those involved in the fabrication of *Museo Aero Solar* can participate in and witness it taking shape, and following that, can witness the event of its taking to the air (should prevailing weather conditions allow). When it takes to the air, *Museo Aero Solar* exemplifies what, following Jill Bennett, we can understand as the "practical aesthetics" of the elemental.[57] Through its form, one of the signature envelopes of the Anthropocene— the plastic bag—is turned into another envelope that takes to the air fueled by nothing but the radiation from the sun.[58] Even if it does not take to the air, however, *Museo Aero Solar* offers compelling experiences of envelopment on the ground. It can be inflated with air, allowing people to enter and move around a space illuminated and colored by the plastic bags and the various advertising and logos printed on them.

Museo Aero Solar draws together diverse constituencies in the affective work of fabricating an envelope that becomes an atmospheric thing; one that can be grasped as an ongoing compositional arrangement of bodies, fabrics, and elements as matters of shared concern.[59] In doing so it recalls and remobilizes the allure of early balloon launches and their capacity to capture attention. But where those early experiments generated public spectacles while keeping science a private affair, the kind of fashioning, assembly, and eventual launch that informs the work of Saraceno is premised on the possibility of generating atmospheric publics in which expertise is distributed—by "atmospheric publics" I mean publics in which the elemental conditions of the meteorological atmosphere become a distributed matter of concern through the generation of a palpable, felt, affective atmosphere of involvement. These publics can be intensified, and given volume, through the shape of the envelope as the form of what Irigaray calls "elemental jouissance."[60] Rather than an envelope that creates a public that closes in on itself, however, the architecture of works such as *Museo Aero Solar* points to the role that envelopment might play in exposing different kinds of bodies to the elemental conditions of their existence. To draw upon Frances Dyson, it suggests not only a form of elementalism involving a relation between the body and its immediate space but also a "permeable body integrated within, and subject to, a global system: one that combines the air we breathe, the weather we feel, the pulses and waves of the electromagnetic spectrum that subtends and

FIGURE 9.2 *Museo Aero Solar*, 21er Haus, Vienna, June 2015. Photo by author.

enables technologies, old and new, and circulates, as Richards would say, in the excitable tissues of the heart."[61]

As atmospheric things taking shape through envelopment, the works in which Saraceno is involved in creating are speculative experiments for an elemental commons. They foreground the importance of cultivating capacities to sense the properties of this commons, while also highlighting the necessity of distributing and sharing devices and technologies through which this sensing takes place. Critically, they remind us that even as non-human sensing and sounding devices of all kinds become more integral to making atmospheres explicit, this does not lead inevitably to the disappearance or dissolution of the body as a locus of elemental sensing. Instead, the key question becomes how to distribute capacities for sensing the elemental across different bodies and forms of life through practices and processes of partial enclosure, generating forms that are open and exposed to the elements. If, as John Durham Peters argues, we "need a better name for the infrastructural aesthetics and ethics of being alive in the cosmos," then piloting speculative concepts like *Aerocene* might furnish important waypoints.[62] This is because they point to the possibility of inventing new arrangements of elemental infrastructure and craft, each of which has the potential to "create its own world afresh."[63] They are reminders that while the act of releasing something into the air can never be taken for granted, on each occasion of its enactment it has the potential to draw out novel arrangements in our collective capacities for sensing and becoming attuned to variations in an elemental commons.

Even if it is increasingly regulated, restricted, and in some case proscribed, there is a value to the act of release that should not and cannot be dismissed: if the circumstances are right this act might produce occasions for generating a necessary collective captivation with the conditions of atmospheres. By generating atmospheres of immersion, such occasions can sometimes work to produce experiences of involvement around important matters of political, social, or cultural concern. They can redirect our attention, not only upward, to the sky, but also towards the very properties and qualities of the elemental milieus in which bodies are generatively implicated. They invite us to think about what it might mean to release something into the air to drift as if it were a gift rather than something that diminished the capacities of the air to sustain different forms of life.

This is what makes a focus on envelopment as a technical process even more important: practices and processes of envelopment, whether in the form of Serres's boxes, Sloterdijk's spheres, or Saraceno's solar sculptures, are important means of making what Peter Adey has called the "force of the elemental" explicit and actionable in distinctive ways.[64] These apprehensions of the elemental are only partial, of course. Nor, indeed, is envelopment the only process or practice through which elemental atmospheres can be disclosed and made explicit, or through which they can become the domain of experiment. And it is worth remembering that the elemental as environmental milieu is by no means reducible to atmosphere. An expansive elementalism would necessarily disclose what Nigel Clark calls "inhuman natures" that link the geological with the atmospheric and with the cosmological by tracing the machines and devices that mediate these phases of matter.[65]

Nevertheless, as part of a renewed elementalism, focusing on envelopment provides an important way in which the politics of atmospheres can be made explicit and experimented with. A politics of elemental envelopment turns around the cofabrication of collectives and associations that disclose variations in the conditions in which different forms of life take shape. This politics takes shape in ways that are simultaneously affective, meteorological, and infrastructural, albeit with different emphases and intensities that are variously palpable and withdrawn. It speculates critically and creatively with elemental envelopment in order to pursue opportunities for thinking about and potentially reworking understandings and apprehensions of atmospheres and the atmospheric. Certainly, this politics discloses how atmospheres become the object-target of various forms of intervention. It problematizes how atmospheres are engineered and designed to affect bodies in different ways, moving them, agitating them, or rendering them still or inert. It encourages wariness of efforts to generate new spaces of alluring commodification in the guise of benign forms of atmospheric togetherness. It recognizes that the condition of envelopment is alluring in all kinds of ways, and has a force felt differentially in bodies—human and nonhuman—as variously wondrous and disquieting, uplifting and disturbing.

A politics of elemental envelopment explores how envelopment provides a technical process through which the elemental forces of atmospheres are differentially mobilized, and to ends that are often far more malign than others: in some instances envelopment is implicated in the distribution of weapons or of unsettling ideas and affects. The process of being and becoming enveloped can be as diminishing as it can be enlivening, as unsettling as comforting. As Andreas Philippopoulos-Mihalopoulos has shown so convincingly, the politics of both atmospheres and envelopment are as much about capacities to avoid enclosure and immersion as they are about generating spaces of open-ended experience and experiment.[66] They are about the possibilities of maintaining and generating different capacities for being enveloped and exposed to the force of the elemental along a spectrum that ranges from the softness of a barely perceptible stir in the air to the overwhelming hardness of agencies and affects that threaten to destroy the conditions for different forms of life. For instance, in December 2015, city officials in Beijing issued a "red alert" because the smog enveloping the city had become dangerously toxic, intensifying to what some English-speaking residents of that city called "airpocalypse." As part of the city's emergency air-pollution response system, the "red alert" triggered a series of restrictions on transportation, school opening hours, and commercial energy use.[67] In this condition, the capacity to reduce exposure to the affects and effects of atmospheres is a defining characteristic of the politics of urban life.[68]

Smog, dust, and clouds of pollutants can also be understood as particular kinds of atmospheric thing. They are far less aesthetically alluring, of course, than many of those that have figured here. Clearly, the atmospheric things that take shape under the name of projects like *Aerocene* are important not because they offer solutions to the kinds of experiences sketched briefly above. But they might provide means by which the kinds of conditions that generate such suffocating atmospheres are disclosed in new ways. And they might allow us to re-imagine and re-appropriate atmospheres as political spacetimes. At the very least, through logics of drift, tentative letting go, and small acts of collective crafting, they remind us that atmospheric spacetimes are certainly not reducible to the macropolitics of territorial containment. Nor, indeed, are they adequately described in terms of the volumetric, even if this term is important because it gives three-dimensional depth and verticality to political spaces. Instead, in certain circumstances, artistic experiments with releasing things into the air remind us that atmospheres are micropolitical spacetimes in motion. Furthermore, through forms of technical expertise and speculative rigor, these spacetimes can become infrastructural conditions for the generation and distribution of multiple forms and experiences of elemental envelopment.[69] These kinds of experiments hold in shape the vital but tensed promise of elemental envelopment with a proposition that draws us in: if the elemental force of atmospheres is never singular, there are many more alluring atmospheric things, perhaps yet to be fabricated, through which to experiment with new sources of value through acts of release, drift, and dispersal. And yet the openness of these things to their elemental conditions is also a reminder that they never close in on themselves in the shape of reassuring containers or settled spheres of immersion. Their power lies in their capacity to draw us out of our envelopes of experience, allowing us, in the process, to sense and feel anew the elemental conditions of atmospheric life.

ANGELS

There is a risk, admittedly, that I am asking too much of this thing, asking it to carry too much weight, when its levity is the property that can potentially do so much. And I realize, even now, that attention to such relatively simple acts and devices in the turbulence of the present can easily appear anachronistic, and for all kinds of reasons. Indeed, for some commentators, critical attention to the atmospheric and elemental qualities of contemporary life means trading in a "focus on technical objects for a concern with the sensory processes to which technics give access."[70] And in some accounts, the body as a locus of intervention and actionability is dissolving into this atmospheric milieu, which is becoming the condition, in turn, for any form of critical intervention or thinking. Frances Dyson, for instance, writes, "The body has given way to the atmosphere—the resonant, information-filled atmosphere—as the site for technological development."[71] At the same time, to foreground simple technologies like the balloon might seem naive in a world where dirigibility, targeting, and persistent predictable presence are virtues. In a world where the drone has become the figure of new technical and political assemblages of security and surveillance, attention to the form, capacities, and promise of less so-phisticated airborne craft might appear wayward at the least.[72]

A qualified focus on such simple devices can however provide oppor-tunities for thinking about atmospheres—about how they are disclosed, sensed, and imagined. And it can provide opportunities for speculating about what is at stake—philosophically, ethically, and politically—when the atmospheric is invoked as the force of something excessive of bodies, enti-ties, or objects. Perhaps this is because, as a device for doing atmospheric things, the balloon is a particularly alluring kind of elemental figure—a figure variously light, heavy, elusive, or inescapable, a figure that draws us in with the promise of sensing a force excessive of its form or of ours. An elemental figure, the very shape of which reminds us of the processuality of form, of how form is a passing between other states of matter.

Other elemental figures also do this, of course. The angel is one. To make this association is to invoke without naively affirming the vision of transcendence that links the act of balloon release with the heavenly incarnations of the figure of the angel; nor is it to make a literal connec-tion between this figure and the blue orbs that feature in Bradbury's *Mar-tian Chronicles*.[73] It is also to think about how both figures help us grasp the relation between elementality, movement, and envelopment. This sense of the angel animates the thinking of two philosophers—Luce Irigaray and Michel Serres—whose work has ventilated the arguments of this book. For both Irigaray and Serres, the angel is a mobile figure with the capacity to move between forms in the process of distributing messages. In *An Ethics of Sexual Difference*, Irigaray outlines a vision of the angel as a fig-ure that "unceasingly *passes through . . . envelope(s) or container(s)*."[74] For Irigaray, angels are "messengers, who transgress all enclosures in their speed, tell of the passage between the envelope of God and that of the world as micro- or macrocosm."[75] Irigaray mobilizes this figure negatively because it seems to act as a guardian preventing man from acknowledging his indebtedness to the maternal envelope of the womb; at the same time, it allows her to imbue the corporeal with the sensible transcendence of the divine.[76] We might read the divine, like Irigaray, as an elemental spirit. But we might also read it as the elemental force of the atmospheric, a force with which the figure of the angel moves, and which moves through us in the passing of breathing as the coming and going of the world.

Similarly, in *Angels: A Modern Myth,* Michel Serres offers a legend of sorts for a mobile world by returning to the figure of the angel as it has been depicted in western art, religion, and literature. As Serres notes, the angel, embodied prototypically in the figure of Mercury, is a messenger. And as messengers, angels are intermediaries "passing among others who are also intermediaries," always in transit, never fully there.[77] If the value of Irigaray's invocation of the angel is to remind us of the danger of assuming the form of the envelope as something closed in on itself, Serres encourages us to attend to how contemporary angels take shape as devices for traveling through and translating the mingled given of an elemental milieu composed of different materials and messages, hardware and software. In this context, angelic messages and messengers are incarnated in many different forms: Serres points, for instance, to the transmission "through the air" of "people, aircraft and electronic signals" and of the various things (letters, voice, data, and so on) they convey.[78] Whether in physical or electromagnetic form, all of these messengers can be seen, according to Serres, as angels: "Angels of steel, carrying angels of flesh and blood, who in turn send angel signals across air waves."[79] The figures of "messenger angels" can also be less technical, taking the form of intermediary "fluxes of nature": a breeze, the rippling of water, or the "heat and light of sun and stars."[80] The wind is one such messenger angel, whose variable force is felt on the skin of bodies of different kinds, human and nonhuman, free and tethered. It can also, of course, become a kind of demon or devil, according to its strength, ferocity, and unpredictability.

For Serres the force of these messages is irreducible to their significance or decipherability for human life. When, for instance, a tree is blown by a gust of wind, it is in some sense translating the message borne by that wind. Equally, albeit far more slowly, a river might be said to translate the message of tectonic uplift as it cuts a channel through the rock. This kind of translating of subtle messages has for the most part been beyond human habits of thinking and thresholds of perception. But for Serres it is a vital part of the wider project of making sense of what he calls "biogea."[81] It requires developing a form of elemental intelligence for understanding how different message-bearing "systems" interact, how transport networks and technologies of circulation, meteorological fluxes, and legendary stories of divine figures all become implicated in ways that give shape and sense to the elemental commons we—humans and nonhumans—inhabit.

As an elemental messenger angel, the balloon is by no means the only device through which to develop this kind of intelligence. And it may not be the best. And yet, however modest, the balloon is a messenger angel for sensing and making explicit the elemental force of "environments and small differences" in ways that make atmospheres more present but never fully.[82] As a messenger angel, the balloon reminds us that the elemental is the always mingled given of which Michel Serres writes, and envelopment is the condition of being in an already mixed milieu. It is a device or craft for mediating, moving, and translating between different atmospheric message-bearing systems. In the shape of piloted balloons it translates elemental variations in the meteorological atmosphere into senses of movement and stillness through its own body and into those of its passengers. In the shape of the radiosonde it translates the dynamics and properties of this atmosphere into data. In many artistic experiments, the balloon is the shape of an envelope that allows for something to pass between the movements of the air, the movements of human bodies, and the movements of voluminous spacetimes. And in small acts of grief and memory, the balloon becomes the shape of our finite exposure to infinitude.

In all of this, the balloon remains an elemental figure, one that mixes and passes between atmosphere and entity. A figure that is never really an envelope closed in on itself but the sensed shape of elemental envelopment in process.

And it is in this shape, then, that I can now perhaps begin to let go.

INTRODUCTION

1. Mischer's words are transcribed from a video of the convention, "Balloon Drop Failure at 2004 DNC—CNN Snafu," YouTube, July 2, 2006, http://www .youtube.com/watch?v=9FTtlbkTWTY, last accessed October 24, 2016.

2. Keating Holland and Adam Levy, "Deflating: Democrats Not into Balloon Drop Tradition," CNN Political Ticker, September 6, 2012, http://politicalticker.blogs .cnn.com/2012/09/06/deflating-democrats-not-into-balloon-drop-tradition/, last accessed October 24, 2016.

3. Treb Heining quoted in Caroline Winter, "The Balloon's Big Moment," Bloomberg Business Week, September 5, 2012, http://www.businessweek.com/articles /2012–09–05/the-balloons-big-moment, last accessed October 24, 2016.

4. For a discussion of Kerry's affective capacities, see Lauren Berlant, "Unfeeling Kerry," Theory and Event 8, no. 2 (2005): n.p.

5. Lichtgrenze website at http://lichtgrenze.de, last accessed December 20, 2016. Details about how this event was planned can be found at http://www.berlin .de/mauerfall2014/en/highlights/balloon-event/index.html, last accessed December 20, 2016.

6. "Balloon-Powered Internet for Everyone," Project Loon, http://www.google .com/loon/how/, last accessed October 24, 2016.

7. "Balloon-Powered Internet for Everyone."

8. I borrow the term "extrusive" from Tim Ingold, The Life of Lines (London: Routledge, 2015).

9. Peter Sloterdijk, Bubbles: Spheres, vol. 1, trans. Wieland Hoban (New York: Semiotext(e), 2011).

10. Michel Serres, The Five Senses: A Philosophy of Mingled Bodies, trans. Margaret Sankey and Peter Cowley (London: Athlone, 2008); Luce Irigaray, An Ethics of Sexual Difference, trans. Carolyn Burke and Gillian C. Gill (London: Continuum, 2005).

11. See the essays in Tim Ingold, The Perception of the Environment (London: Routledge, 2000); Tim Ingold, Making: Anthropology, Art, and Architecture (London: Routledge, 2013); and Bettina Hauge, "The Air from Outside: Getting to Know the World through Air Practices," Journal of Material Culture 18, no. 2 (2013): 171–87.

12. James Ash, The Interface Envelope: Gaming, Technology, Power (London: Bloomsbury, 2015).

13. Bernard Stiegler, *Technics and Time: The Fault of Epimetheus*, vol. 1, trans. Richard Beardsworth and George Collins (Stanford, CA: Stanford University Press, 1998).

14. On enchantment, see Jane Bennett, *The Enchantment of Modern Life: Attachments, Crossings, Ethics* (Princeton, NJ: Princeton University Press, 2001).

15. For instance, see Gernot Böhme, "Atmosphere as the Fundamental Concept of a New Aesthetics," *Thesis Eleven* 36 (1993): 113–26; and Tonino Griffero, *Atmospheres: Aesthetics of Emotional Spaces* (Farnham, UK: Ashgate, 2014).

16. On this, see Ben Anderson, *Encountering Affect: Capacities, Apparatuses, Conditions* (Farnham, UK: Ashgate, 2014); Lauren Berlant, *Cruel Optimism* (Durham, NC: Duke University Press, 2011); Teresa Brennan, *The Transmission of Affect* (Ithaca, NY: Cornell University Press, 2004); and Kathleen Stewart, "Atmospheric Attunements," *Environment and Planning D: Society and Space* 29, no. 3 (2011): 445–53.

17. Ingold, *Life of Lines*. See also Derek McCormack, "Engineering Affective Atmospheres on the Moving Geographies of the 1897 Andrée Expedition," *Cultural Geographies* 15, no. 4 (2008): 413–30; and Peter Adey, "Air/Atmospheres of the Megacity," *Theory, Culture, and Society* 30, nos. 7–8 (2013): 291–308.

18. Stewart, "Atmospheric Attunements"; and Ben Anderson and John Wylie, "On Geography and Materiality," *Environment and Planning A* 41, no. 2 (2009): 318–35.

19. Graham Harman, "Realism without Materialism," *SubStance* 40 (2011): 64; Graham Harman, *Guerrilla Metaphysics: Phenomenology and the Carpentry of Things* (Peru, IL: Open Court, 2010); and Timothy Morton, *Hyperobjects: Philosophy and Ecology after the End of the World* (Minneapolis: University of Minnesota Press, 2013).

20. Jane Bennett, *Vibrant Matter: A Political Ecology of Things* (Durham, NC: Duke University Press, 2010); Jane Bennett, "Systems and Things: A Response to Graham Harman and Timothy Morton," *New Literary History* 43, no. 2 (2012): 227.

21. Ingold, *Life of Lines*, 16. Ingold rails against the deadening effects of OOO as a philosophical position, making a series of telling points. For the sake of balance, a more affirmative account of the promise of OOO is offered by Timothy Morton, "Here Comes Everything: The Promise of Object-Oriented Ontology," *Qui Parle: Critical Humanities and Social Sciences* 19, no. 2 (2011): 163–90.

22. Stewart, "Atmospheric Attunements."

23. Timothy Choy and Jerry Zee, "Condition: Suspension," *Cultural Anthropology* 30, no. 2 (2015): 211.

24. The most acidic critique of these terms comes from Timothy Morton, who points out its importance in various kinds of "ecomimetic" nature writing. Timothy Morton, *Ecology without Nature: Rethinking Environmental Aesthetics* (Cambridge, MA: Harvard University Press, 2007). For a more tempered critique of the term, see Andreas Philippopoulos-Mihalopoulos, *Spatial Justice: Body, Lawscape, Atmosphere* (London: Routledge, 2015).

25. Hans Ulrich Gumbrecht, *Atmosphere, Mood, Stimmung: On a Hidden Potential of Literature*, trans. Erik Butler (Stanford, CA: Stanford University Press, 2012), 12.

26. Here I extend important arguments about landscape made by cultural geographers such as John Wylie and Mitch Rose. John Wylie, "Landscape, Absence and the Geographies of Love," *Transactions of the Institute of British Geographers* 34, no. 3 (2009): 275–89; and Mitch Rose, "Gathering Dreams of Presence: A Project for the Cultural Landscape," *Environment and Planning D: Society and Space* 24, no. 4 (2006): 537–54.

27. On the geography of immersive experience, see Harriet Hawkins and Elizabeth Straughan, "Nano-Art, Dynamic Matter, and the Sight/Sound of Touch," *Geoforum* 51, no. 1 (2014): 130–39.

28. Peter Sloterdijk, "Airquakes," *Environment and Planning D: Society and Space* 27, no. 1 (2009): 41 57.

29. Philippopoulos-Mihalopoulos, *Spatial Justice*. Also Anderson, *Encountering Affect*; Peter Adey, "Security Atmospheres or the Crystallization of Worlds," *Environment and Planning D: Society and Space* 32, no. 5 (2014): 834–51; Marjin Nieuwenhuis, "Atmospheric Governance: Gassing as law for the protection and killing of life," *Environment and Planning D: Society and Space*, forthcoming, https://doi.org/10.1177/0263775817729378

30. By "forms of life" I mean patterned assemblages of associations between humans and nonhumans.

31. Timothy Choy, *Ecologies of Comparison: An Ethnography of Endangerment in Hong Kong* (Durham, NC: Duke University Press, 2011); Sasha Engelmann, "Toward a Poetics of Air: Sequencing and Surfacing Breath," *Transactions of the Institute of British Geographers* 40, no. 3 (2015): 430–44.

32. Stewart, "Atmospheric Attunements."

33. Derek McCormack, "Devices for Doing Atmospheric Things," in *Nonrepresentational Methodologies: Re-envisioning Research*, ed. Phillip Vannini, 89–111 (London: Routledge, 2015).

34. Kathleen Stewart, *Ordinary Affects* (Durham, NC: Duke University Press, 2007).

35. Jamie Lorimer, Timothy Hodgetts, and Maan Barua, "Animals' Atmospheres," *Progress in Human Geography* (forthcoming).

36. Ash, *Interface Envelope*; Mark Hansen, *Feed-Forward: On the Future of Twenty-First-Century Media* (Chicago, IL: University of Chicago Press, 2014).

37. Ingold, *Life of Lines*; and Bronislaw Szerszynski, "Reading and Writing the Weather: Climate Technics and the Moment of Responsibility," *Theory, Culture, and Society* 27 nos. 2–3 (2010): 9–30.

38. Ian Bogost, *Alien Phenomenology, or, What It's Like to Be a Thing* (Minneapolis: University of Minnesota Press, 2012).

39. Ian Cook, "Follow the Thing: Papaya," *Antipode* 36, no. 4 (2004): 642–64.

40. Kathleen Stewart, "Studying Unformed Objects: The Provocation of a Compositional Mode," *Fieldsights—Field Notes, Cultural Anthropology Online*,

June 30, 2013, http://culanth.org/fieldsights/350-studying-unformed-objects-the -provocation-of-a-compositional-mode, last accessed September 5, 2017.

41. Derek McCormack, *Refrains for Moving Bodies: Experience and Experiment in Affective Spaces* (Durham, NC: Duke University Press, 2013).

42. See, instead, Richard Holmes, *Falling Upwards: How We Took to the Air* (London: William Collins, 2013); Michael Lynn, *The Sublime Invention: Ballooning in Europe, 1783–1820* (London: Pickering and Chatto, 2010); and Peter Haining, ed., *The Dream Machines: An Eye-Witness History of Ballooning* (London: New English Library, 1972).

43. Donna Haraway, "Situated Knowledges: The Science Question in Feminism and the Privilege of Partial Perspective," *Feminist Studies* 14, no. 3 (1988): 575–99.

44. Michel Serres, "Jules Verne's Strange Journeys," *Yale French Studies* 52 (1975): 177.

45. Laura Salisbury, "Michel Serres: Science, Fiction, and the Shape of Relation," *Science Fiction Studies* 33, no. 1 (2006): 31.

46. Stewart, "Atmospheric Attunements," 447.

47. Bennett, *Vibrant Matter*.

48. Harman, *Guerrilla Metaphysics*.

CHAPTER 1. ENVELOPMENT

1. Donald Barthelme, *Sixty Stories* (New York: Penguin, 2003).

2. Barthelme, *Sixty Stories*, 46.

3. Barthelme, *Sixty Stories*, 46.

4. Barthelme, *Sixty Stories*, 47.

5. Barthelme, *Sixty Stories*, 48.

6. Barthelme, *Sixty Stories*, 51.

7. Philip Childs, *Texts: Contemporary Cultural Texts and Critical Approaches* (Edinburgh, Scotland: Edinburgh University Press, 2006).

8. Barthelme, *Sixty Stories*, 47.

9. Barthelme, *Sixty Stories*, 49.

10. To some extent this interest is bound up in the wider affective turn in the social sciences and humanities, as scholars in disciplines such as anthropology, cultural studies, geography, and sociology work to develop vocabularies for grasp- ing the form of affective spacetimes. See Ben Anderson, "Affective Atmospheres," *Emotion, Space, and Society* 2, no. 2, (2009): 77–81; James Ash, "Rethinking Affective Atmospheres: Technology, Perturbation and Space-Times of the Non-Human," *Geo- forum* 49, no. 1 (2013): 20–28; Lauren Berlant, *Cruel Optimism* (Durham, NC: Duke University Press, 2011); Mikkel Bille, "Lighting Up Cosy Atmospheres in Denmark," *Emotion, Space, and Society* 15, no. 1 (2015): 56–63; and David Bissell, "Passenger Mobilities: Affective Atmospheres and the Sociality of Public Transport," *Environment and Planning D: Society and Space* 28, no. 2 (2010): 270–89.

11. Tim Edensor and Shanti Sumartajo, "Designing Atmospheres: Introduction to Special Issue," *Visual Communication* 14, no. 3 (2015): 251–65; Mikkel Bille, Peter

Bjerregard, and Tim Flohr Sørensen, "Staging Atmospheres: Materiality, Culture, and the Texture of the In-Between," *Emotion, Space, and Society* 15, no. 1 (2015): 31–38; and Cameron Duff, "Atmospheres of Recovery and Assemblages of Health," *Environment and Planning A*, 48, no. 1 (2016): 58–74.

12. Ben Anderson, *Encountering Affect: Capacities, Apparatuses, Conditions* (Farnham, UK: Ashgate, 2014).

13. Anderson, "Affective Atmospheres."

14. Such understandings of atmospheres have been shaped by a phenomenological tradition of thinking about atmospheric aesthetics, exemplified in the work of Gernot Böhme, for whom atmospheres are the primary object of aesthetic perception. Gernot Böhme, "Atmosphere as the Fundamental Concept of a New Aesthetics," *Thesis Eleven* 36 (1993): 113–26. For a discussion of the phenomenology of atmospheres, see Tonino Griffero, *Atmospheres: Aesthetics of Emotional Spaces* (Farnham, UK: Ashgate, 2014). Böhme's phenomenological exploration of the aesthetics of atmosphere is paralleled by a related strand of research into ambience, a concept discussed much earlier by Leo Spitzer. Leo Spitzer, "Milieu and Ambience: An Essay in Historical Semantics," *International Phenomenological Society* 3, no. 2 (1942): 169–218. Recent research in ambience is exemplified in the work of Jean-Paul Thibaud. For Thibaud, ambience figures as the background feel of environmental experience, or as the world considered from "a sensory point of view." Jean-Paul Thibaud, "The Sensory Fabric of Urban Ambiances," *Senses and Society* 6, no. 2 (2011): 203–15. For a discussion of the resonances between research into atmosphere and ambience, see Peter Adey, Laure Brayer, Damien Masson, Patrick Murphy, Paul Simpson, and Nicholas Tixier, "'Pour votre tranquilité': Ambiance, Atmosphere, and Surveillance," *Geoforum* 49 (2013): 299–309. It is worth remarking upon the difference between these terms. "Ambience," for scholars like Thibaud, is the world considered from a sensory point of view. In that sense, and as others have noted, there may be many overlaps between the concepts "atmosphere" and "ambience." However, "atmosphere" may be preferable for two reasons. First, the concept speaks more directly to gaseous senses of spacetime in ways that foreground the relation between forms of life and the atmospheric conditions in which they are immersed. Second, "ambience" seems to suggest a relatively tranquil, and to some extent benign, affective spacetime. "Atmospheres," in contrast, can be threatening, turbulent, and volatile in ways that are simply not suggested by "ambience." However, and this is an important point, atmospheres can also be ambient: that is, they are distributed, diffuse envelopes of affective materiality in which bodies are immersed in a relation of generative conditioning.

15. Teresa Brennan, *The Transmission of Affect* (Ithaca, NY: Cornell University Press, 2004).

16. Anderson, "Affective Atmospheres," 80.

17. Anderson, *Encountering Affect*.

18. Anderson, *Encountering Affect*; and Peter Adey, "Security Atmospheres or the Crystallization of Worlds," *Environment and Planning D: Society and Space* 32, no. 5 (2014): 834–51.

19. Tim Ingold, *Being Alive* (London: Routledge, 2015); and Peter Adey, "Air/Atmospheres of the Megacity," *Theory, Culture, and Society* 30, nos. 7–8 (2013): 291–308.

20. For discussions of different ways in which the meteorological atmosphere is framed politically, see, for example, Timothy Choy, *Ecologies of Comparison: An Ethnography of Endangerment in Hong Kong* (Durham, NC: Duke University Press, 2011); Julie Cupples, "Culture, Nature and Particulate Matter—Hybrid Reframings in Air Pollution Scholarship," *Atmospheric Environment* 43, no. 1 (2009): 207–17; and Mark Whitehead, *State, Science, and the Skies: Environmental Governmentality and the British Atmosphere* (Oxford: Wiley-Blackwell, 2009).

21. Brian Massumi, *Ontopower: War, Powers, and the State of Perception* (Durham, NC: Duke University Press, 2015); and Joseph Masco, *The Theatre of Operations: National Security Affect from the Cold War to the War on Terror* (Durham, NC: Duke University Press, 2014).

22. Bronislaw Szerszynski, "Reading and Writing the Weather: Climate Technics and the Moment of Responsibility," *Theory, Culture, and Society* 27, nos. 2–3 (2010): 9–30.

23. Ingold, *Being Alive*.

24. Mark Jackson and Maria Fannin, "Letting Geography Fall Where It May—Aerographies Address the Elemental," *Environment and Planning D: Society and Space* 29, no. 3 (2011): 435–44; Craig Martin, "Fog-Bound: Aerial Space and the Elemental Entanglements of Body-with-World," *Environment and Planning D: Society and Space* 29, no. 3 (2011): 454.

25. Ben Anderson and John Wylie, "On Geography and Materiality," *Environment and Planning A* 41, no. 2 (2009): 318–35.

26. On the concept of perturbation, see Levi Bryant, *Onto-Cartography: An Ontology of Machines and Media* (Edinburgh, Scotland: Edinburgh University Press, 2014).

27. Alfred North Whitehead, *The Concept of Nature* (New York: Prometheus Books, 2004).

28. Mark Hansen, *Feed-Forward: On the Future of Twenty-First-Century Media* (Chicago, IL: University of Chicago Press, 2014); Malcolm McCullough, *Ambient Commons: Attention in the Age of Embodied Information* (Cambridge, MA: MIT Press, 2013); and Ulrik Schmidt, "Ambience and Ubiquity," in *Throughout: Art and Culture Emerging with Ubiquitous Computing*, ed. Ulrik Ekman, 175–88 (Cambridge, MA: MIT Press, 2013).

29. For critiques of the tendency for some media scholars to forget the infrastructures that sustain media, see John Durham Peters, *The Marvelous Clouds: A Philosophy of Elemental Media* (Chicago, IL: University of Chicago Press, 2015); and Nicole Starosielski, *The Undersea Network* (Durham, NC: Duke University Press, 2015).

30. Alfred North Whitehead, *Process and Reality*, corrected ed. (New York: Free Press, 1927). Whitehead does use the term "entity," or "actual entity," albeit in a way that is interchangeable with "actual occasion."

31. Jane Bennett, *Vibrant Matter: A Political Ecology of Things* (Durham, NC: Duke University Press, 2010), 20.

32. Barthelme, *Sixty Stories*, 50.

33. For an overview, see Penny Harvey et al., eds., *Objects and Materials: A Routledge Companion* (London: Routledge, 2014).

34. For an overview of different approaches here, see Graham Harman, *Towards Speculative Realism: Essays and Lectures* (Ropley, UK: Zero Books, 2010); Peter Gratton, *Speculative Realism: Prospects and Problems* (London: Bloomsbury, 2014); and Steven Shaviro, *The Universe of Things: On Speculative Realism* (Minneapolis: University of Minnesota Press, 2014).

35. Timothy Morton "Here Comes Everything: The Promise of Object-Oriented Ontology," *Qui Parle: Critical Humanities and Social Sciences* 19, no. 2 (2011): 163–90.

36. Graham Harman, "Realism without Materialism," *SubStance* 40 (2011): 64.

37. Harman, "Towards Speculative Realism," 199.

38. Graham Harman, "Whitehead and Schools X, Y, and Z," in *The Lure of Whitehead*, ed. Nicholas Gaskill and A. J. Nocek (Minneapolis: University of Minnesota Press, 2014), 231–48. It is also worth noting here that Harman distinguishes between philosophies of process and becoming, and identifies Whitehead's work in terms of the former, Deleuze's in terms of the latter, and his own in terms of neither.

39. A. N. Whitehead, *Process and Reality*.

40. A. N. Whitehead, *Nature*, 77.

41. A. N. Whitehead, *Nature*, 78.

42. William Connolly, *A World of Becoming* (Durham, NC: Duke University Press, 2011); Erin Manning, *Relationscapes: Movement, Art, Philosophy* (Cambridge, MA: MIT Press, 2009); Tom Roberts, "From Things to Events: Whitehead and the Materiality of Process," *Environment and Planning D: Society and Space* 32, no. 6 (2014): 968–83; and Nicholas Gaskill and A. J. Nocek, eds., *The Lure of Whitehead* (Minneapolis: University of Minnesota Press, 2014).

43. To be fair, this is clear not only from Barthelme's story but also from other versions of speculative realism in which the argument for the importance of an entity-centric account of reality fails to escape the necessity of positing something between entities, whether this is "mesh," in the case of Timothy Morton, or "world," in the case of Levi Bryant. Timothy Morton, *Hyperobjects: Philosophy and Ecology at the End of the World* (Minneapolis: University of Minnesota Press, 2013); and Bryant, *Onto-Cartography*.

44. For a much broader and detailed discussion of the relation between Whitehead's work and speculative realism, see Shaviro, *Universe of Things*.

45. Harman, "Realism without Materialism," 177.

46. Harman, of course, is not alone in making this claim. Timothy Morton, for instance, makes a similarly strong claim when he argues for the importance of hyperobjects as entities that are massively distributed in spacetime. In making this claim, Morton argues against any account of reality that relies upon a kind of worldly substance in which these entities are immersed. Morton, *Hyperobjects*.

47. Timothy Morton, *Ecology without Nature: Rethinking Environmental Aesthetics* (Cambridge, MA: Harvard University Press, 2007), 89.

48. Morton, *Hyperobjects*.

49. Morton, *Hyperobjects*.

50. Indeed, because of this, there is arguably far more sympathy between the kind of hauntology that emerges from Derrida's account of spectrality and Harman's account of allure than the latter might admit. Read through Derrida, "allure" is perhaps another word for the spectral quality of things.

51. On atmosphere as condition and object-target, see Anderson, *Encountering Affect*.

52. Ben Anderson and James Ash, "Atmospheric Methods," in *Nonrepresentational Methods: Re-envisioning Research*, ed. Phillip Vannini, 34–51 (London: Routledge, 2015).

53. Anderson, "Affective Atmospheres"; Anderson and Ash, "Atmospheric Methods."

54. Ingold, *Being Alive*.

55. Cymene Howe and Dominic Boyer, "Aeolian Politics," *Distinktion: Scandinavian Journal of Social Theory* 16, no. 1 (2015): 31.

56. Brian Massumi, *Parables for the Virtual: Movement, Affect, Sensation* (Durham, NC: Duke University Press, 2002).

57. Peter Hallward, *Out of This World: Deleuze and the Philosophy of Creation* (London: Verso, 2006), 38.

58. Michel Serres, *Genesis*, trans. Genevieve James and James Nielson (Ann Arbor: University of Michigan Press, 1995), 112.

59. Serres, *Genesis*, 111.

60. Serres, *Genesis*, 103.

61. Serres, *Genesis*, 110. On the relation between Serres's understanding of turbulence and the materiality of spacetimes, see Tim Cresswell and Craig Martin, "On Turbulence: Entanglements of Disorder and Order on a Devon Beach," *Tijdschrift Voor Economische en Sociale Geographie* 103 (2012): 516–29.

62. Anderson and Wylie, "On Geography and Materiality"; Jane Bennett and William Connolly, "The Crumpled Handkerchief," in *Time and History in Deleuze and Serres*, ed. Bernd Herzogenrath, 153–72 (London: Continuum, 2012).

63. Bennett and Connolly, "Crumpled Handkerchief."

64. Serres, *Genesis*, 102.

65. Bennett, *Vibrant Matter*.

66. Jane Bennett, "Systems and Things: A Response to Graham Harman and Timothy Morton," *New Literary History* 43, no. 2 (2012): 227.

67. Gilles Deleuze, *The Fold*, trans. Tom Conley (London: Athlone, 1993).

68. David Giessen, *Manhattan Atmospheres: Architecture, the Interior Environment, and Urban Crisis* (Minneapolis: University of Minnesota Press, 2013). While architecture is an obvious example of the importance of such envelopment, it can be traced in other ways. Consider, for instance, the kinds of packaging

and gases that makes possible the protective atmospheres in which much of the food in industrial societies is enveloped, and to the air bags that have now become essential parts of cars as mobile envelopes of security and safety in everyday life.

69. Peter Sloterdijk, *In the World Interior of Capital*, trans. Wieland Hoban (Cambridge: Polity Press, 2013).

70. Serres, *Genesis*.

71. Serres, *Genesis*, 45.

72. L. Chollet, "Balloon Fabrics Made of Goldbeater's Skins," *L'aéronautique*, August 1922, 5, http://www.gasballooning.net/Fabric%20Goldbeater%20Skin.pdf, last accessed October 20, 2014.

73. Alan Guth, *The Inflationary Universe: The Quest for a New Theory of Cosmic Origins* (London: Jonathan Cape, 1997).

74. Henri Bergson, *The Creative Mind* (Mineola, NY: Dover, 2007), 77.

75. For an overview, see Michael Lynn, *The Sublime Invention: Ballooning in Europe, 1783–1820* (London: Pickering and Chatto, 2010).

76. Luce Irigaray, *An Ethics of Sexual Difference*, trans. Carolyn Burke and Gillian C. Gill (London: Continuum, 2005).

77. Irigaray, *Sexual Difference*; and Luce Irigaray, *The Forgetting of Air in Martin Heidegger*, trans. Mary Beth Mader (London: Athlone, 1999). For a discussion of the relation between surfaces and spacing using the insights of Irigaray, see Rachel Colls and Maria Fannin, "Placental Surfaces and the Geographies of Bodily Interiors," *Environment and Planning A* 45, no. 5 (2013): 1087–104.

78. For a critique of the tension between Irigaray's elemental materialism and her ethics of sexual difference, see Alison Stone, "Irigaray's Ecological Phenomenology: Towards an Elemental Materialism," *Journal of the British Society for Phenomenology* 46, no. 2 (2015): 117–31. For Stone, "Irigaray's strong claim about dual sexuate worlds is actually in tension with her elemental materialism, which supports a more moderate view of how far sexuation affects perceptual experience. On that view, sexuate differences qualify perception but do not divide it radically, because shared perceptual structures—such as the role of breath in mediating how we inhabit the shared world—result from men's and women's bodily constitution from the same elemental matters. We can therefore embrace Irigaray's elemental materialism without having to endorse her problematic belief in dual sexuate worlds. Once extricated from that problematic belief, Irigaray's materialism deserves to be developed further, for it does important work in rethinking basic ontological issues in a way that addresses the ecological crisis" (130).

79. Irigaray, *Sexual Difference*, 9.

80. For a discussion of how Irigaray's work can inform understandings of architecture, see Peg Rawes, *Irigaray for Architects* (London: Routledge, 2007).

81. Michel Serres, *The Five Senses: A Philosophy of Mingled Bodies*, trans. Margaret Sankey and Peter Cowley (London: Athlone, 2008), 302.

82. On the concept of "thing-power," see Bennett, *Vibrant Matter*.

1. "Ballon de Paris," http://www.ballondeparis.com, last accessed May 5, 2014.

2. Derek McCormack, "Remotely Sensing Affective Afterlives: The Spectral Geographies of Material Remains," *Annals of the Association of American Geographers* 100, no. 3 (2010): 640–54; Derek McCormack, "Engineering Affective Atmospheres on the Moving Geographies of the 1897 Andrée Expedition," *Cultural Geographies* 15, no. 4 (2008): 413–30.

3. Gunter Söllinger, *S. A. Andrée: The Beginning of Polar Aviation, 1895–1897* (Moscow: Russian Academy of Sciences, 2005).

4. "The Big Balloon in Paris," *New York Times*, 5 August 1878.

5. L. T. C. Rolt, *The Balloonists: A History of the First Aeronauts* (Stroud, UK: Sutton, 2006).

6. Mark Dorrian and Frédéric Pousin, eds., *Seeing from Above: The Aerial View in Visual Culture* (London: I. B. Tauris, 2013).

7. "Big Balloon in Paris."

8. Tethered balloons also informed the visual experience of other attractions at these events. Patrick Ellis, "'Panoramic Whew': The Aeroscope at the Panama-Pacific International Exposition, 1915," *Early Modern Popular Visual Culture* 13, no. 3 (2015): 209–31.

9. John Wylie, "An Essay on Ascending Glastonbury Tor," *Geoforum* 33, no. 4 (2002): 445.

10. I draw the idea of "site-conditioned" from the work Robert Irwin. Robert Irwin, *Being and Circumstance: Notes towards a Conditional Art* (San Francisco, CA: Lapis Press, 1985).

11. Gaston Bachelard, *Air and Dreams: An Essay on the Imagination of Movement*, trans. Edith R. Farrell and C. Frederick Farrell (Dallas, TX: Dallas Institute Publications, 1988).

12. Caren Kaplan, "The Balloon Prospect," in *From Above: War, Violence, and Verticality*, ed. Peter Adey, Mark Whitehead, and Alison Williams, 19–40 (London: Hurst, 2013); and *Aerial Aftermaths: Wartime from Above* (Durham: Duke University Press, 2018). See also Anders Ekström, "Seeing from Above: A Particular History of the General Observer," *Nineteenth-Century Contexts* 31, no. 3 (2009): 185–207; and Richard Hallion, *Taking Flight: Inventing the Aerial Age, from Antiquity through the First World War* (Oxford: Oxford University Press, 2003).

13. James Glaisher quoted in Elaine Freedgood, *Victorian Writing about Risk: Imagining a Safe England in a Dangerous World* (Cambridge: Cambridge University Press, 2000), 89.

14. James Campbell, *Introduction to Remote Sensing*, 4th ed. (London: Taylor Francis, 2006), 6.

15. John Pickles, ed., *Ground Truth: The Social Implications of GIS* (New York: Guildford, 1995); John Pickles, *A History of Spaces: Cartographic Reason, Mapping and the Geo-Coded World* (London: Routledge, 2003); and Denis Cosgrove,

A Cartographic Genealogy of the Earth in the Western Imagination (Baltimore, MD: Johns Hopkins University Press, 2001).

16. Richard Beck, "Remote Sensing and GIS as Counterterrorism Tools in the Afghanistan War: A Case Study of the Zhawar Kili Region," *Professional Geographer* 55, no. 2 (2003): 170–79; Jody Berland, "Remote Sensors: Canada and Space," *Semiotext(e)* 6, no. 2 (1994): 28–25.

17. Dorrian and Pousin, *Seeing from Above.*

18. Gillian Rose, *Feminism and Geography: On the Limits of Geographical Knowledge* (Minneapolis: University of Minnesota Press, 1993); Donna Haraway, *Simians, Cyborgs, and Women: The Reinvention of Nature* (London: Routledge, 1991); and Donna Haraway, "The Persistence of Vision," in *The Visual Culture Reader*, ed. Nicholas Mirzoeff (London: Routledge, 2001), 191–98.

19. James Glaisher quoted in Rolt, *Balloonists*, 83.

20. Gilles Deleuze, *Expressionism in Philosophy: Spinoza*, trans. Martin Joughin (New York: Zone Books, 1990), 127.

21. Wylie, "Glastonbury Tor."

22. Kaplan, "Balloon Prospect."

23. Nadar (Gaspard-Félix Tournachon), "My Life as Photographer," trans. Thomas Repensek, *October* 5 (1978): 6–28. See also Paula Amad, "From God's Eye to Camera-Eye: Aerial Photography's Post-Humanist and Neo-Humanist Visions of the World," *History of Photography* 36, no. 1 (2012): 66–86.

24. David Clarke and Marcus Doel, "Engineering Space and Time: Moving Pictures and Motionless Trips," *Journal of Historical Geography* 31 no. 1 (2005): 41–60.

25. John Wylie, "A Single Day's Walking: Narrating Self and Landscape on the Southwest Coast Path," *Transactions of the Institute of British Geographers* 30, no. 2 (2005): 234–47.

26. Camille Flammarion quoted in James Glaisher, ed., *Travels in the Air*, 2nd ed. (London: Richard Bentley and Son, 1871), 147.

27. Martyn Barber and Helen Wickstead, "'One Immense Black Spot': Aerial Views of London, 1784–1918," *London Journal* 35, no. 3 (2010): 236–54.

28. Alfred North Whitehead, *Process and Reality*, corrected ed. (New York: Free Press, 1927).

29. Nathan Brown, "The Technics of Prehension: On the Photography of Nicholas Baier," in *The Lure of Whitehead*, ed. Nicholas Gaskill and A. J. Nocek (Minneapolis: University of Minnesota Press, 2014), 143. Following the writing of Levi Bryant, we might think of the sensing of atmospheres by this envelope as a form of perturbation; the balloon is perturbed by the circumstantial motion and variation of the field in which it moves, circumstances that are merely the localized expression of a much larger and much wider distributed medium called atmosphere. Levi Bryant, *Onto-Cartography: An Ontology of Machines and Media* (Edinburgh, Scotland: Edinburgh University Press, 2014).

30. Graham Harman, "Whitehead and Schools X, Y, and Z," in *The Lure of Whitehead*, ed. Nicholas Gaskill and A. J. Nocek (Minneapolis: University of Minnesota Press, 2014), 231–48.

31. Swedish Society for Anthropology and Geography, *The Andrée Diaries: Being the Diaries and Records of S. A. Andrée, Nils Strindberg, and Knut Fraenkel, Written during Their Balloon Expedition to the North Pole in 1897 and Discovered on White Island in 1930, Together with a Complete Record of the Expedition and Discovery*, trans. E. Adams-Ray (London: Bodley Head, 1931), 48.

32. Monck Mason, *Aeronautica; Or, Sketches Illustrative of the Theory and Practice of Aerostation: Comprising an Enlarged Account of the Late Aerial Expedition to Germany* (London: F. C. Westley, 1838), 132.

33. Glaisher, *Travels in the Air*, 2.

34. Jennifer Tucker, "Voyages of Discovery on Oceans of Air: Scientific Observation in an Age of 'Balloonacy,'" *Osiris* 11 (2006): 144–76.

35. Mason, *Aeronautica*, 119.

36. Mason, *Aeronautica*, 116–17.

37. Mason, *Aeronautica*, 135.

38. Mason, *Aeronautica*, 129.

39. Albert Santos-Dumont quoted in Philip Hoffman, *Wings of Madness: Albert Santos-Dumont and the Invention of Flight* (New York: Theia, 2003), 41.

40. David Bissell, "Animating Suspension: Waiting for Mobilities," *Mobilities* 2, no. 2 (2007): 227–98.

41. For a parallel account of sensing in the elemental medium of water, see Elizabeth Straughan, "Touched by Water: The Body in Scuba Diving," *Emotion, Space, and Society* 5, no. 1 (2012): 19–26.

42. Freedgood, *Victorian Writing*.

43. Monck Mason quoted in Freedgood, *Victorian Writing*, 78.

44. Wilfred de Fonvielle quoted in Glaisher, *Travels in the Air*, 284.

45. Glaisher, *Travels in the Air*, 94.

46. Santos-Dumont quoted in Hoffman, *Wings of Madness*, 41.

47. Bachelard, *Air and Dreams*, 27.

48. Archimedes, *On Floating Bodies I*, in T. L. Heath, *The Works of Archimedes* (Cambridge: Cambridge University Press, 1897), 257.

49. Michel Serres, *Biogea*, trans. Randolph Burks (Minneapolis, MN: Univocal Publishing, 2012), 64.

50. Serres, *Biogea*, 65.

51. Vojtech Jirat-Wasiutynski, "The Balloon as Metaphor in the Early Work of Odilon Redon," *Artibus et Historiae* 13 (1992): 195–206.

52. Émile Hennequin quoted in Alfred Werner, *Introduction to the Graphic Works of Odilon Redon* (New York: Dover Publications, 2005), 7.

53. Odilon Redon quoted in Werner, *Introduction*, 8.

54. Mitch Rose, "Gathering Dreams of Presence: A Project for the Cultural Landscape," *Environment and Planning D: Society and Space* 24, no. 4 (2006): 537–54.

55. Quoted in Campbell, *Remote Sensing*, 6.

56. Jacques Derrida, *Specters of Marx: The State of the Debt, the Work of Mourning, and the New International*, trans. Peggy Kamuf (New York: Routledge, 1994).

57. A. N. Whitehead, *Process and Reality*, 121.

58. Rose, "Gathering Dreams of Presence."

59. John Wylie, "The Spectral Geographies of W. G. Sebald," *Cultural Geographies* 14, no. 2 (2007): 171–88.

60. Gilles Deleuze, *Difference and Repetition*, trans. Paul Patton (London: Continuum, 1994), 139.

61. Erin Manning, *Politics of Touch: Movement, Sense, Sovereignty* (Minneapolis: University of Minnesota Press, 2007).

62. On an expanded array of contemporary technologies and practices of sensing, see Jennifer Gabrys, *Environmental Sensing Technology and the Making of a Computational Planet* (Minneapolis: University of Minnesota Press, 2016).

CHAPTER 3. ALLURE

1. Graham Harman, *Guerrilla Metaphysics: Phenomenology and the Carpentry of Things* (Peru, IL: Open Court, 2010). For a discussion of the wider implications of this argument, see the chapters in Roland Faber and Andrew Goffey, eds., *The Allure of Things* (London: Bloomsbury, 2014). The relation between allure and other ways of thinking about the affective dimensions of nonhumans is also worth thinking about further, but is beyond the scope of my concerns here. See Jamie Lorimer, "Nonhuman charisma," *Environment and Planning D: Society and Space* 25, no. 5 (2007): 911–32.

2. Steven Shaviro, *Post-Cinematic Affect* (Ropely, UK: O-Books, 2010), 9.

3. Steven Shaviro, *The Universe of Things: On Speculative Realism* (Minneapolis: University of Minnesota Press, 2014).

4. Nigel Thrift, "Understanding the Material Practices of Glamour," in *The Affect Theory Reader*, ed. Melissa Gregg and Gregory Seigworth (Durham, NC: Duke University Press, 2010), 296. For a discussion of the allure of surfaces more generally, see Isla Forsyth, Hayden Lorimer, Peter Merriman, and James Robinson, "What Are Surfaces?" *Environment and Planning A* 45, no. 5 (2013): 1013–20.

5. Fraser MacDonald, "Perpendicular Sublime: Regarding Rocketry and the Cold War," in *Observant States: Geopolitics and Visual Cultures*, ed. Fraser MacDonald, Rachel Hughes, and Klaus Dodds, 267–90 (London: I. B. Tauris, 2010).

6. Benjamin Franklin to Sir Joseph Banks, August 30, 1783, http://archive.org /stream/benjaminfranklin00rotciala/benjaminfranklin00rotciala_djvu.txt, last accessed on July 2, 2014.

7. Paul Keen, "The 'Balloonomania': Science and Spectacle in 1780s England," *Eighteenth-Century Studies* 39, no. 4 (2006): 507–53.

8. Michael Lynn, *The Sublime Invention: Ballooning in Europe, 1783–1820* (London: Pickering and Chatto, 2010).

9. Thrift, "Glamour," 290.

10. Mi Gyung Kim, "Balloon Mania: News in the Air," *Endeavour* 28, no. 4 (2004): 149–55.

11. Lynn, *Sublime Invention*, 78.

12. Quoted in Vincenzo Lunardi, *An Account of the First Aerial Voyage in England: In a Series of Letters to His Guardian, Chevalier Gherardo Compagni, Written under the Impressions of the Various Events That Affected the Undertaking* (London: Bell Publishers, 1784), 72. For an example of how to tell stories through attending to the qualities of "things," see Caitlin DeSilvey, "Observed Decay: Telling Stories with Mutable Things," *Journal of Material Culture* 11, no. 3 (2006): 318–38.

13. L. T. C Rolt, *The Balloonists: The History of the First Aeronauts* (Stroud, UK: Sutton, 2006), 69.

14. Lunardi, *Account*, 31.

15. Marc Dessauce, *The Inflatable Moment: Pneumatics and Protest in '68* (New York: Princeton Architectural Press, 1999); Richard Holmes, *Falling Upwards: How We Took to the Air* (London: William Collins, 2013); Sean Topham, *Blow Up: Inflatable Art, Architecture, and Design* (London: Prestel, 2002).

16. This included the use of the view from above within new forms of visual experience. See Erkki Huhtamo, "Aeronautikon! Or, The Journey of the Balloon Panorama," *Early Popular Visual Culture* 7, no. 3 (2009): 295–306. On the relation between the balloon and theater, see Kate Turner, "The Spectacle of Democracy in the Balloon Plays of the Revolutionary Period," *Forum for Modern Language Studies* 39, no. 3 (2003): 241–53; and John Robbins, "Up in the Air: Balloonomania and Scientific Performance," *Eighteenth-Century Studies* 48, no. 4 (2015): 521–38.

17. Mark Twain, *Tom Sawyer Abroad, and Tom Sawyer, Detective* (London: Wordsworth, 2009), 15–16.

18. As John Tresch notes, Poe's balloon story "reproduces the principles of the technological news article, with the content suggested by the public's fascination with mechanical wonders and details supplied by various pamphlets and encyclopaedic sources." John Tresch, "'The Potent Magic of Verisimilitude': Edgar Allan Poe within the Mechanical Age," *British Journal for the History of Science* 30 (1997): 283.

19. On April 15, 1844, a retraction appeared in the *Sun* acknowledging that the story may have been erroneous, but adding that "the description of the Balloon and the voyage was written with a minuteness and scientific ability calculated to obtain credit everywhere, and was read with great pleasure and satisfaction. We by no means think such a project impossible." The Edgar Allan Poe Society of Baltimore, http://www.eapoe.org/works/tales/ballhxa.htm, last accessed March 20, 2013.

20. John Tully, *The Devil's Milk: A Social History of Rubber* (New York: Monthly Review Press, 2011).

21. Frank Hoffman, Frederick Augustyn, and Martin Manning, *Dictionary of Toys and Games in American Popular Culture* (New York: Routledge, 2004).

22. Tully, *Devil's Milk*, 43.

23. E. E. Cummings, *Selected Poems* (New York: Norton, 1994), 4.

24. S. Harpal Singh, "From Hunters to Balloon Sellers," *Hindu*, April 19, 2010, http://www.hindu.com/2010/04/19/stories/2010041954190200.htm, last accessed November 3, 2015.

25. James Ash, "Technologies of Captivation: Videogames and the Attunement of Affect," *Body and Society* 19, no. 1 (2013): 27–51.

26. William Frank Browne, "Improvement in Balloon Advertising," US Patent Number 144436A, published November 11, 1873.

27. Robert Wilson, "Advertising-Balloon," US Patent Number 496177A, published April 25, 1893.

28. Arthur Salzer, "Illuminated Toy Balloon and Lighting Effect," US Patent Number 1229794A, published June 12, 1917.

29. J. Maguire Eustace, "Balloon Advertising Method and Apparatus," US Patent Number 1598211A, published August 31, 1926.

30. Stephen Harp, *Marketing Michelin: Advertising and Cultural Identity in Twentieth-Century France* (Baltimore: Johns Hopkins University Press, 2001).

31. Keith A. Sculle and John A. Jakle, "Signs in Motion: A Dynamic Agent in Landscape and Place," *Journal of Cultural Geography* 25, no. 1 (2008): 57–85.

32. The parade has run every year with the exception of the years from 1942 to 1944, when rubber was scarce because of the war effort. In the earlier years, balloons were released, with their finders rewarded with cash prizes and trophies. "16 Get Balloon Prizes: Fall River Boy Receives Macy's Silver Trophy," *New York Times*, December 27, 1930, 30. The larger of these balloons were designed with leaks to give them a duration in the air of about a week. It made financial sense to offer rewards, as the larger balloons were both expensive and reusable. "100,000 Children See Store Pageant: Macy's Christmas Parade Is Greeted by Cheers along Broadway," *New York Times*, November 30, 1928, 25.

33. Thrift, "Glamour." Guliana Bruno, *Surface: Matters of Aesthetics, Materiality, and Media* (Chicago, IL: University of Chicago Press, 2014); James Ash, "Rethinking Affective Atmospheres: Technology, Perturbation and Space-Times of the Non-Human," *Geoforum* 49, no. 1 (2013): 20–28.

34. Thrift, "Glamour," 300.

35. "What's It to You?," Film Department, E. I. DuPont Denemours, 1955.

36. Mimi Sheller, *Aluminum Dreams: The Making of Light Modernity* (Cambridge, MA: MIT Press, 2014).

37. Peter Sloterdijk, *In the World Interior of Capital: For a Philosophical Theory of Globalization*, trans. Wieland Hoban (Cambridge: Polity Press, 2013).

38. Lothar Bucher quoted in Marshall Berman, *All That Is Solid Melts into Air: The Experience of Modernity* (London: Verso, 2010), 239–40.

39. Rayner Banham, "Monumental Wind Bags," in *The Inflatable Moment: Pneumatics and Protest in '68*, ed. Marc Dessauce, 31–33 (New York: Princeton Architectural Press, 1999).

40. Dessauce, *Inflatable Moment*, 126.

41. Jeffrey Meikle, *American Plastic: A Cultural History* (New Brunswick, NJ: Rutgers University Press, 1997).

42. Dessauce, *Inflatable Moment*, 29.

43. Georges Teyssot, "Responsive Envelopes: The Fabric of Climatic Islands," *Appareil* 11 (2013): n.p.

44. Billy Klüver, ed., *Pavilion: Experiments in Art and Technology* (New York: Dutton, 1972).

45. On the more general relation between atmospheres and consumption, see Brigitte Biehl-Missal and Michael Saren, "Atmospheres of Seduction: A Critique of Aesthetic Marketing Practices," *Journal of Macromarketing* 32, no. 2 (2012): 168–80; and Stephen Healy, "Atmospheres of Consumption: Shopping as Involuntary Vulnerability," *Emotion, Space, and Society* 10, no. 1 (2014): 35–43. On the importance of the relation between ambient power and the seduction of spaces, see John Allen, "Ambient Power: Berlin's Potsdamer Platz and the Seductive Logic of Public Spaces," *Urban Studies* 43, no. 2 (2006): 441–55.

46. Pernilla Ohrstedt, "Feeling Reflective," *RIBJA Newsletter*, March 1, 2013, https://www.ribaj.com/culture/feeling-reflective, last accessed December 20, 2016.

47. Randall Packer, "The Pavilion: Into the 21st Century: A Space for Reflection," *Organized Sound* 9, no. 3 (2004): 253; and Billy Klüver, "Photographic Recording of Some Optical Effects in a 27.5m Spherical Mirror," *Applied Optics* 10, no. 12 (1971): 27–55.

48. Documented in the film by Eero Saarinen. EAT, "The Great Big Mirror Dome Project," *Experiments in Art and Technology Los Angeles Records, 1969–1979* (Los Angeles: The John Paul Getty Trust, 1970).

49. Packer, "Pavilion," 252.

50. David Erdman, Marcelyn Gow, Ulrika Karlsson, and Chris Perry, "Parallel Processing: Design/Practice," *Architectural Design* 76 (2006): 81.

51. Teyssot, "Responsive Envelopes."

52. Fred Turner, "The Corporation and the Counterculture: Revisiting the Pepsi Pavilion and the Politics of Cold War Multimedia," *Velvet Light Trap* 73 (2014): 66–78. On the politics of domes as immersive media spaces, see David McConville, "Cosmological Cinema: Pedagogy, Propaganda, and Perturbation in Early Dome Theatres," *Technoetic Arts* 5, no. 2 (2007): 69–85.

53. McConville, "Cosmological Cinema," 68.

54. McConville, "Cosmological Cinema," 76.

55. Mark Hansen, *Feed-Forward: On the Future of Twenty-First-Century Media* (Chicago, IL: University of Chicago Press, 2014).

56. Paul Miller, "The Engineer as Catalyst: Billy Klüver on Working with Artists," *Spectrum, IEEE* 35, no. 7 (1998): 20–29, 25.

57. David Joselit, "Yippie Pop: Abbie Hoffman, Andy Warhol, and Sixties Media Politics," *Grey Room* 8 (2002): 75.

58. Miller, "Engineer as Catalyst," 25.

59. Merce Cunningham quoted in Judith Mackrell, "The Joy of Sets," *Guardian*, June 6, 2005, http://www.theguardian.com/stage/2005/jun/06/dance, last accessed November 16, 2015.

60. Marcia Siegel, "Dancing in the Dust," *Hudson Review* 54, no. 3 (2001): 463.

61. Siegel, "Dancing in the Dust," 463.

62. Siegel, "Dancing in the Dust," 463.

63. For more on the Zygote Balls, see Tangible Interaction's website at http://www.tangibleinteraction.com, last accessed November 3, 2015. See also http://www.tangibleinteraction.com/rentals, last accessed September 6, 2017.

64. This work was developed further in *Open Burble* (2006), in which a similar cloud was released into the air as part of the Singapore Biennale, and in *Burble London* (2007). *Sky Ear* at http://www.haque.co.uk/skyear.php, last accessed November 4, 2015.

65. Janet Echelman, *Skies Painted with Unnumbered Sparks*, http://www.echelman.com/project/skies-painted-with-unnumbered-sparks/, last accessed March 26, 2014.

66. Janet Echelman and Aaron Koblin, *Unnumbered Sparks*, http://www.unnumberedsparks.com, last accessed March 26, 2014.

67. Ash, "Rethinking Affective Atmospheres."

68. To focus on the latter is to rehearse a metaphysical claim about a kind of essential withdrawal that can easily fall back on a kind of speculative mysticism, and at the expense of attention to how the qualities of allure are actively fabricated and engineered to particular ends.

69. Jane Bennett, *The Enchantment of Modern Life: Attachments, Crossings, Ethics* (Princeton, NJ: Princeton University Press, 2001).

70. Patricia Clough, "The Transformational Object of Cruel Optimism," 2011, http://bcrw.barnard.edu/wp-content/uploads/2012/Public-Feelings-Responses/Patricia-Ticineto-Clough-The-Transformational-Object-of-Cruel-Optimism.pdf, last accessed 3 November 2015.

71. Bennett, *Enchantment*.

72. Nigel Thrift, "Glamour," 296.

CHAPTER 4. RELEASE

1. Directed by Michael Anderson and written by Richard Matheson, *The Martian Chronicles* TV series (1980) was based on Ray Bradbury's series of stories of the same name.

2. From *The Martian Chronicles* TV series, near the end of the second episode.

3. Ray Bradbury, *The Martian Chronicles* (London: Harper Collins, 2008), 114.

4. Ray Bradbury, "Take Me Home," *New Yorker*, June 4, 2012, http://www.newyorker.com/reporting/2012/06/04/120604fa_fact_bradbury, last accessed June 19, 2014.

5. Bradbury, *Martian Chronicles*, 154–55.

6. Elizabeth Bishop, *Poems* (New York: Farrar, Straus and Giroux, 2011), 101.

7. At the outset of this chapter it is worth noting that my aim here is not to develop a detailed discussion of the environmental ethics of balloon release or of its connection to other forms of life.

8. John Wylie, "Landscape, Absence and the Geographies of Love," *Transactions of the Institute of British Geographers* 34, no. 3 (2009): 280.

9. Wylie, "Geographies of Love," 281.

10. For other ways of telling the stories of the geographies of love, see Carey-Ann Morrison, Lynda Johnston, and Robyn Longhurst, "Critical Geographies of Love as Spatial, Relational and Political," *Progress in Human Geography* 37, no. 4 (2013): 505–21.

11. Wylie, "Geographies of Love," 284.

12. Jane Bennett, *Vibrant Matter: A Political Ecology of Things* (Durham, NC: Duke University Press, 2010).

13. Jean-Luc Nancy, *The Inoperative Community*, ed. Peter Connor, trans. Peter Connor, Lisa Garbus, Michael Holland, and Sinoma Sawhney (Minneapolis: University of Minnesota Press, 1991).

14. Judith Butler, *Precarious Life: The Powers of Mourning and Violence* (London: Verso, 2004), 23.

15. Butler, *Precarious Life*, 23.

16. Dana Luciano, *Arranging Grief: Sacred Time and the Body in Nineteenth-Century America* (New York: New York Press, 2007), 2.

17. Luciano, *Arranging Grief*, 20.

18. Butler, *Precarious Life*, 39.

19. Elizabeth Hallam and Jenny Hockey, *Death, Memory, and Material Culture* (Oxford: Berg, 2001).

20. Peggy Phelan, *Unmarked: The Politics of Performance* (London: Routledge, 1993).

21. Sara Ahmed, "Happy Objects," in *The Affect Theory Reader*, ed. Melissa Gregg and Gregory Seigworth (Durham, NC: Duke University Press, 2010), 29–51.

22. Ahmed, "Happy Objects."

23. "Funeral or Remembrance Balloon Release," Balloons for U, http://www.balloonsforu.co.uk/funeral-balloon-releases/, last accessed June 18, 2014.

24. Reagan Library Archives, "Proclamation 5890—Pregnancy and Infant Loss Awareness Month, 1988," October 25, 1988, https://www.reaganlibrary.archives.gov/archives/speeches/1988/102588b.htm, last accessed September 07, 2017.

25. Hope Mommies blog at http://hopemommies.org/balloon-release, last accessed June 18, 2014.

26. A Legacy of Love website at http://alegacyofloveloss.wordpress.com/2013/09/26/3rd-annual-balloon-release-mayors-park-visalia-ca-october-15-2013-530pm/, last accessed June 18, 2014.

27. Gathered from a web search undertaken on March 20, 2013.

28. Patricia Clough, "Praying and Playing to the Beat of a Child's Metronome," *Subjectivity* 3 (2010): 349–65.

29. "Balloon Released in Memory of Karina Menzies Found in Germany," *Wales Online*, http://www.walesonline.co.uk/news/wales-news/2012/11/01/balloon-released-in-memory-of-karina-menzies-found-in-germany-91466-32140879/#ixzz2OBWTEGBB, last accessed March 20, 2013.

30. Kate Flynn, "Fat and the Land: Size Stereotyping in Pixar's *Up*," *Children's Literature Association Quarterly* 35, no. 4 (2010): 435–42; Casey Brienza, "Remem-

bering the Future: Cartooning Alternative Life Courses in *Up* and *Future Lovers*," *Journal of Popular Culture* 46 (2013): 299–314.

31. Jules Verne, *Around the World in Eighty Days, and Five Weeks in a Balloon* (London: Wordsworth Classics, 1994).

32. Gilles Deleuze, *Immanence, Essays on a life*, trans. Anne Boyman (New York: Zone Books, 2001).

33. Lauren Berlant quoted in Luciano, *Arranging Grief*, 235.

34. Julian Barnes, *Levels of Life* (London: Jonathan Cape, 2013), 10, 12.

35. Barnes, *Levels of Life*, 37.

36. Barnes, *Levels of Life*, 84.

37. Barnes, *Levels of Life*, 103.

38. Barnes, *Levels of Life*, 116.

39. Ian McEwan, *Enduring Love* (London: Jonathan Cape, 1997), 1.

40. McEwan, *Enduring Love*, 13.

41. Barnes, *Levels of Life*, 67.

42. James Williams, "Love in a Time of Events: Badiou, Deleuze and White-head on Chesil Beach," draft paper, https://whiteheadresearch.org/occasions/conferences/event-and-decision/papers/James%20Williams_Final%20Draft.pdf, last accessed September 07, 2017; McEwan, *Enduring Love*, 1.

43. McEwan, *Enduring Love*, 3.

44. McEwan, *Enduring Love*, 3.

45. Gilles Deleuze and Félix Guattari, *A Thousand Plateaus*, trans. Brian Massumi (London: Athlone, 1987).

46. Geoffrey Bennington, *Interrupting Derrida* (London: Routledge, 2000), 1. Michel Serres, *The Five Senses: A Philosophy of Mingled Bodies*, trans. Margaret Sankey and Peter Cowley (London: Continuum, 2008), 293.

47. McEwan, *Enduring Love*, 31.

48. McEwan, *Enduring Love*, 31–32.

49. McEwan, *Enduring Love*, 32.

50. McEwan, *Enduring Love*, 32.

51. McEwan, *Enduring Love*, 2.

52. McEwan, *Enduring Love*, 3.

53. Serres, *Five Senses*.

54. Rachel Jones, "Vital Matters and Generative Materiality: Between Bennett and Irigaray," *Journal of the British Society for Phenomenology* 46, no. 22 (2015): 156–72.

55. Paul Harrison, "*Flētum*: A Prayer for X," *Area* 43 (2011): 159.

56. Harrison, "*Flētum*," 160.

57. Harrison, "*Flētum*," 160.

58. Harrison, "*Flētum*," 160.

59. Andrew Metcalfe, "Nothing: The Hole, the Holy, the Whole," *Ecumene* 8, no. 3 (2001): 247–63.

1. Through observing what happened to hydrogen in his balloons, and through careful experiment with the behavior of gas in inverted glass containers, Charles derived a formulation that has since become a standard in chemistry and physics: simply put, it states that at constant pressure the volume of a given mass of an ideal gas is inversely proportional to its temperature. While the law is named after him, Charles did not publish it. Almost fifteen years later, Joseph Louis Gay-Lussac, also a balloonist, published a paper that outlined the elements of the law while also acknowledging the earlier work of Charles. Charles Holbrow and Joseph Amato, "What Gay-Lussac Didn't Tell Us," *American Journal of Physics* 79, no. 1 (2011): 17–24.

2. Derek McCormack, "Remotely Sensing Affective Afterlives: The Spectral Geographies of Material Remains," *Annals of the Association of American Geographers* 100, no 3. (2010): 640–54.

3. *Flight of the Eagle (Ingenjör Andrées luftfärd)*, directed by Jan Troel (Sweden: Bold Productions, 1982).

4. Tim Ingold, *The Life of Lines* (London: Routledge, 2015).

5. Peter Sloterdijk, *Bubbles: Spheres*, vol. 1, trans. Wieland Hoban (Los Angeles: Semiotext(e), 2011).

6. Peter Sloterdijk quoted in Jean-Pierre Couture, "Spacing Emancipation? Or How Sphereology Can Be Seen as a Therapy for Modernity," *Environment and Planning D: Society and Space* 27, no. 1 (2009): 162.

7. Nigel Thrift, "Different Atmospheres: Of Sloterdijk, China, and Site," *Environment and Planning D: Society and Space* 27, no. 1 (2009): 119–38; and Nigel Thrift, "Peter Sloterdijk and the Philosopher's Stone," in *Sloterdijk Now*, ed. Stuart Elden (Cambridge: Polity Press, 2012), 133–46.

8. Sloterdijk, *Bubbles*, 28.

9. Peter Sloterdijk, *Terror from the Air*, trans. Amy Patton and Steve Corcoran (Cambridge, MA: MIT Press, 2009).

10. As Stuart Elden has noted, the term "volumetric" was coined by Jeremy Crampton. Stuart Elden, "Secure the Volume: Vertical Geopolitics and the Depth of Power," *Political Geography* 34, no. 1 (2013): 35–51; Jeremy Crampton, "Cartographic Calculations of Territory," *Progress in Human Geography* 35, no. 1 (2010): 92–103.

11. Elden, "Secure the Volume." For a discussion of the politics of verticality and volume in a different context, see Eyal Weizman, *Hollow Land: Israel's Architecture of Occupation* (New York: Verso, 2012).

12. Mark Whitehead, *State, Science, and the Skies: Environmental Governmentality and the British Atmosphere* (Oxford: Wiley-Blackwell, 2009).

13. Henri Lefebvre, *The Production of Space*, trans. D. Nicholson-Smith (Oxford: Blackwell, 1992), 337.

14. Peter Adey, "Securing the Volume/Volumen: Comments on Stuart Elden's Plenary Paper 'Secure the Volume,'" *Political Geography* 34, no. 1 (2013): 52–54.

15. Peter Adey, Laure Brayer, Damien Masson, Patrick Murphy, Paul Simpson, and Nicolas Tixier, "'Pour votre tranquillité': Ambiance, Atmosphere, and Surveillance," *Geoforum* 49, no. 3 (2013): 299–309.

16. Gilles Deleuze and Felix Guattari, *A Thousand Plateaus*, trans. Brian Massumi (London: Athlone, 1987), 488.

17. Sloterdijk, *Bubbles*, 18. Originally called "A Child's World."

18. Sloterdijk, *Bubbles*, 18.

19. Gerald Silk, "Myth and Meanings in Manzoni's Merda d'Artista," *Art Journal* 52, no. 3 (1993): 65–75.

20. Steven Connor, *The Matter of Air: Science and Art of the Ethereal* (London: Reaktion Books, 2010).

21. Sasha Engelmann, "Toward a Poetics of Air: Sequencing and Surfacing Breath," *Transactions of the Institute of British Geographers* 40, no. 3 (2015): 430–44.

22. Timothy Choy, *Ecologies of Comparison: An Ethnography of Endangerment in Hong Kong* (Durham, NC: Duke University Press, 2011).

23. On sensing the distributed geography of breathing, see Michaela Palmer and Owain Jones, "On Breathing and Geography: Explorations of Data Sonifications of Timespace Processes with Illustrating Examples from a Tidally Dynamic Landscape (Severn Estuary, UK)," *Environment and Planning A* 46, no. 1 (2014): 222–40.

24. Found at http://www.saveenergy.vic.gov.au/getthefacts/whatisblackballoon .aspx, last accessed March 15, 2008. Hyperlink no longer active. This ad was part of the "You Have the Power: Save Energy" TV campaign.

25. David Rood, "Black Balloons Inflate Green Awareness," TheAge.com, March 13, 2007, http://www.theage.com.au/news/national/black-balloons-inflate -green-awareness/2007/03/12/1173548107194.html, last accessed December 6, 2013. To some extent the campaign seems to have had some effect: the Victoria state government reported that the number of subscribers to its GreenPower supplier scheme increased from 80,000 in December 2005 to 157,000 in December 2006. On the relation between materiality, devices, and publics, see Noortje Marres, *Material Participation: Technology, the Environment, and Everyday Publics* (Basingstoke, UK: Palgrave Macmillan, 2012).

26. Sloterdijk, *Bubbles*.

27. George Martin Huss, "The Gas-Tank Nuisance," *Art World* 1, no. 1 (1916): 66.

28. Huss, "Gas-Tank Nuisance," 67.

29. In an editorial in July 2011, the British-based newspaper the *Guardian* regretted the removal of one of these gasometers in north London, calling it an "unquestionably beautiful industrial relic." "In Praise of Gasometers," editorial, *Guardian*, July 20, 2011, http://www.theguardian.com/commentisfree/2011/jul/20 /in-praise-of-gasometers, last accessed April 29, 2014.

30. In Vienna the disused city gasworks were transformed into residential buildings with the external skin retained, and the inner space converted into light-filled courtyards. In Reick, an eastern suburb of Dresden, there are also two distinctive gasometers. The larger of the two is a disused empty shell. The smaller structure, built in 1879, has been transformed into a visitor attraction known as the

Panometer, housing a panoramic display of Dresden's urban landscape created by the Austrian artist Yadegar Asisi in 2006. Based upon a series of historical drawings of the city, the panorama is 27 meters high and 105 meters across, and depicts baroque Dresden as it might have appeared around 1756. In some ways, the effect is to produce the kind of aerial view of cities that free or tethered balloons afforded during the universal exhibitions of the nineteenth and early twentieth centuries. To ascend the central observation tower in the Panometer is, in a strange sense, then, to ascend into the air above baroque Dresden in order to consume a visual spectacle while also hearing its sounds.

31. Gasometer Oberhausen, http://www.gasometer.de/en/, last accessed November 5, 2015.

32. Christo and Jeanne-Claude, http://www.christojeanneclaude.net/projects /big-air-package?view=info#.UpR975GDT10, last accessed April 29, 2014. This was not the first such construction but marked the latest in a series of such air packages constructed since 1966 by Matthias Koddenberg, "'It Boggles the Mind': The Air Packages of Christo and Jeanne-Claude," in *Christo: Big Air Package*, Wolfgang Volz and Peter Pachinke (Essen: Klartext, 2013), 131–43.

33. Messer is the largest privately owned industrial gas company in the world.

34. The envelope was not, however, the first to be inflated and displayed in this gasometer. The centerpiece of an exhibition called *Wind of Hope* (2004) was a fifty-five-meter-high balloon called the *Breitling Orbiter III*: a balloon similar to the one with which Bertrand Piccard and Brian Jones had become the first to circumnavigate the globe nonstop in 1999 during a flight that lasted nineteen days. Housed in the Gasometer, the balloon was illuminated by the light-architecture-engineering company Start.Design.

35. Wolfgang Volz and Peter Pachinke, eds., *Christo: Big Air Package* (Essen, Germany: Klartext, 2013).

36. Peter Schneemann, "Monumentalism as a Rhetoric of Impact," *Anglia* 131, nos. 2–3 (2013): 283.

37. Schneemann, "Monumentalism," 285.

38. Schneemann, "Monumentalism," 291.

39. Schneemann, "Monumentalism," 292.

40. Hal Foster, "Polemics, Postmodernism, Immersion, Militarized Space," *Journal of Visual Culture* 3, no. 3 (2004): 329.

41. Michel Serres, *Biogea*, trans. Randolph Burks (Minneapolis, MN: Univocal, 2012), 37.

42. Serres, *Biogea*, 37.

43. Serres, *Biogea*, 38.

44. For Christo, the fabric of the wrapping generates an equally valuable aesthetic experience: "The wrapped object—be it a tree, a building, a tower, a bridge, whatever—can also be beautiful in its packaging." Volz and Pachinke, *Christo*, 72.

45. Michel Serres, *The Five Senses: A Philosophy of Mingled Bodies*, trans. Margaret Sankey and Peter Cowley (London: Athlone, 2008).

46. Jonathan Jones, "Let There Be Light," *Observer*, Sunday, March 19, 2000.

47. Erin Manning, "Propositions for the Verge: William Forsythe's Choreographic Objects," *Inflexions* 2 (2009): n.p.

48. Manning, "Propositions for the Verge."

49. Madeleine Gins and Arakawa, *Architectural Body* (Tuscaloosa: Alabama University Press, 2002), 11.

50. Nigel Thrift, "Understanding the Material Practices of Glamour," in *The Affect Theory Reader*, ed. Melissa Gregg and Gregory Seigworth (Durham, NC: Duke University Press, 2010), 296.

51. Engelmann, "Toward a Poetics of Air."

52. Luce Irigaray quoted in Elizabeth Hirsh and Gary Olsen, "'Je-Luce Irigaray': A Meeting with Luce Irigaray," trans. Elizabeth Hirsh and Gaëtan Brulotte, *JAC* (1996): 346. It is worth noting that Irigaray here is clarifying what she claims is the mistranslation of her argument about volume elsewhere.

CHAPTER 6. SOUNDING

1. For an excellent discussion of the relation between sound, atmosphere, and envelopment, see Frances Dyson, *Sounding New Media: Immersion and Embodiment in the Arts and Culture* (Berkeley: University of California Press, 2009). See also Michael Gallagher, Anja Kanngieser, and Jonathan Prior, "Listening Geographies: Landscape, Affect and Geotechnologies," *Progress in Human Geography* 41, no. 5 (2017): 618–35.

2. In James Glaisher, ed., *Travels in the Air*, 2nd ed. (London: Richard Bentley and Son, 1871), 174.

3. Glaisher, *Travels in the Air*, 79.

4. Glaisher, *Travels in the Air*, 79.

5. Gay-Lussac's law, formulated first in 1802, states that when the volume and mass of a gas remain constant, then the pressure of that gas will increase with temperature. In addition to first recognizing iodine as a distinct element, Gay-Lussac, with Alexander von Humboldt, also observed that the composition of the atmosphere does not vary with altitude. In pursuit of this understanding of the quality of atmosphere and to investigate the strength of the earth's magnetic field at altitude, Gay-Lussac undertook a balloon flight with Jean-Baptiste Biot in August 1804.

6. C. J. S., "A Series of Letters on Acoustics, Addressed to Mr. Alexander, Durham Palace, West Hackney," *Gentleman's Magazine* 111, no. 2 (February 1812): 110.

7. C. J. S., "Series of Letters," 110.

8. Camille Flammarion quoted in Glaisher, *Travels in the Air*, 2.

9. Glaisher, *Travels in the Air*, 128.

10. Glaisher, *Travels in the Air*, 147.

11. Dyson, *Sounding New Media*, 4.

12. Glaisher, *Travels in the Air*, 148. See also Camille Flammarion, *L'atmosphère: Météorologie populaire* (Paris: Hachette, 1888). Flammarion's speculations anticipate later ideas about the limit of an atmospheric envelope of sonic experience and

materiality, including, notably, Antonin Artaud's "There Is No More Firmament," in *Collected Works*, vol. 2, trans. Victor Corti (London: Calder and Boyars, 1971). For a discussion of Artaud's work, see Dyson, *Sounding New Media*.

13. On the relation between sound and the voice, see Anja Kanngieser, "A Sonic Geography of Voice: Towards an Affective Politics," *Progress in Human Geography* 36, no. 3 (2012): 336–53. On the relation between sound, experience, and spatiality, see George Revill, "How Is Space Made in Sound? Spatial Mediation, Critical Phenomenology and the Political Agency of Sound," *Progress in Human Geography* 40, no. 2 (2016): 240–56.

14. Michel Serres, *The Five Senses: A Philosophy of Mingled Bodies*, trans. Margaret Sankey and Peter Cowley (London: Athlone, 2008), 141.

15. Michel Serres, *Genesis*, trans. Genevieve James and James Nielson (Ann Arbor: University of Michigan Press, 1995), 119.

16. *The Big Bounce*, directed by Jerry Fairbanks (United States: Jerry Fairbanks Productions, 1960).

17. I adapt this phrase from the title of Frances Dyson's book. Frances Dyson, *The Tone of Our Times: Sound, Sense, Economy, and Ecology* (Cambridge, MA: MIT Press, 2014).

18. For an account of the role of satellites in the expansion of media worlds, see Lisa Parks, *Cultures in Orbit: Satellites and the Televisual* (Durham, NC: Duke University Press, 2005).

19. "Preliminary Design of an Experimental World-Circling Spaceship," Report No. SM-11827 (Santa Monica, CA: Douglas Aircraft Company, 1946).

20. Merton Davies and William Harris, "RAND's Role in the Evolution of Balloon and Satellite Observation Systems, and Related US Space Technology" (Santa Monica, CA: The RAND Corporation, R-3692-RC, 1988), 16.

21. Arthur C. Clarke, "Extra-Terrestrial Relays: Can Rocket Stations Give World-Wide Radio Coverage?" *Wireless World* 51 (1945): 305–8.

22. Jon Gertner, *The Idea Factory: Bell Labs and the Great Age of American Invention* (London: Penguin, 2012).

23. William Jakes, *Participation of Bell Telephone Laboratories in Project Echo and Experimental Results: Technical Note D-1127* (Washington, DC: NASA, 1961).

24. Jakes, *Participation of Bell Telephone*, 1.

25. "The Odyssey of Project Echo," *SP-4308 Spaceflight Revolution* (Washington, DC: NASA History Office, n.d.), 156.

26. "Odyssey of Project Echo," 176.

27. "Odyssey of Project Echo," 176.

28. Mimi Sheller, *Aluminum Dreams: The Making of Light Modernity* (Cambridge, MA: MIT Press, 2014).

29. "Odyssey of Project Echo," 188.

30. Strictly speaking, the Echo balloon was not the first surface against which voice signals had been bounced. In 1954 the US Navy began experimenting with the use of the moon as a surface against which to reflect voice transmissions. Helen Gavaghan, *Something New Under the Sun: Satellites and the Beginning of the Space*

Race (New York: Springer, 1998). On the relation between space and media, see Katarina Damjanov, "The Matter of Media in Outer Space: Technologies of Cosmobiopolitics," *Environment and Planning D: Society and Space* 33, no. 5 (2015): 889–906.

31. Transcribed from *The Big Bounce*. See also "Odyssey of Project Echo," 188.

32. "Odyssey of Project Echo," 191. John Pierce, "Satellite Science and Technology," *Observatory* 83 (1963): 207–16.

33. "Odyssey of Project Echo," 192.

34. Dyson, *Tone of Our Times*.

35. Joseph Masco, *Theatre of Operations: National Security Affect from the Cold War to the War on Terror* (Durham, NC: Duke University Press, 2014).

36. Davies and Harris, "RAND's Role," 17.

37. Davies and Harris, "RAND's Role," 42.

38. "Odyssey of Project Echo," 176.

39. "Odyssey of Project Echo," 153.

40. "Odyssey of Project Echo," 155.

41. "GHOST Balloon Completes One-Year Flight," *National Center for Atmospheric Research* (Boulder, Colorado), October 1 1968, 1, http://nldr.library.ucar.edu /repository/assets/info/INFO-000-000-000-105.pdf, last accessed June 4, 2014.

42. "GHOST Balloon Completes One-Year Flight," 2.

43. Jean Pierre Pommereau, "Observation Platforms: Balloons," in *Encyclopedia of Atmospheric Sciences*, ed. J. Holton, J. Pyle, and J. A. Curry (Waltham, MA: Academic Press, 2003), 1429–38.

44. Vincent Lally Papers, University Corporation for Atmospheric Research/National Center for Atmospheric Research, Boulder, Colorado.

45. Lally Papers.

46. "GHOST Balloon Completes One-Year Flight," 2.

47. Lally Papers.

48. Lally Papers.

49. Jussi Parikka, *A Geology of Media* (Minneapolis: University of Minnesota Press, 2015), 12–13.

50. For a discussion of the fraught, complex, and contested "discovery" of cosmic rays, see Charles Ziegler, "Technology and the Process of Scientific Discovery: The Case of Cosmic Rays," *Technology and Culture* 30, no. 4 (1989): 939–63.

51. Qiaozhen Xu and Laurie Brown, "The Early History of Cosmic Ray Research," *American Journal of Physics* 55, no. 23 (1987): 23–32.

52. Gavaghan, *Something New Under the Sun*.

53. Stephen Webb, *Measuring the Universe: Cosmological Distance Ladder* (New York: Springer, 1999), 270.

54. Geoff Brumfiel, "Big Bang's Ripples: Two Scientists Recall Their Big Discovery," *NPR*, May 20, 2014, http://www.npr.org/2014/05/20/314239930/big-bangs -afterglow-two-scientists-recall-their-big-discovery, last accessed August 14, 2015.

55. Arno Penzias, transcribed from "The Violent Universe: The Holmdel Horn Antenna," *BBC* Films, http://www.bbc.co.uk/science/space/universe/scientists /arno_penzias_robert_wilson#poobf13x, last accessed August 14, 2015.

56. Arno Penzias and Robert Wilson, "A Measurement of Excess Antenna Temperature at 4080 Mc/s," *Astrophysical Journal* 142 (1965): 419–21.

57. Timothy Morton, *Hyperobjects: Philosophy and Ecology after the End of the World* (Minneapolis: University of Minnesota Press, 2013).

58. On this point I differ from Timothy Morton's depiction of an "ambient poetics," which he claims is characteristic of ecomimesis. While I agree that the "ambient" can refer to something "material and physical, though somewhat intangible," I don't think we need to inevitably think of it in terms of "world." Nor am I so dismissive of various aesthetic practices, including those that figure in this book, that experiment with the possibility of some sense of a surround. Timothy Morton, *Ecology without Nature: Rethinking Environmental Aesthetics* (Cambridge, MA: Harvard University Press, 2007), 33.

59. On ambience as the felt quality of space from a phenomenological point of view, see Jean-Paul Thibaud, "The Sensory Fabric of Urban Ambiances," *Senses and Society* 6, no. 2 (2011): 203–15.

60. Peter Coles, *The Routledge Critical Dictionary of the New Cosmology* (New York: Routledge, 1999), 244.

61. Serres, *Genesis*.

62. Bruce Bassett, Bob Nichol, and Daniel Eisenstein, "Sounding the Dark Cosmos," *Astronomy and Geophysics* 46, no. 5 (2005): 526–29.

63. P. de Bernardis et al. "Images of the Early Universe from the BOOMERanG Experiment." In *Relativistic Astrophysics*. AIP Conference Proceedings. No. 586, ed. J. C. Wheeler and H. Martel (Melville, NY: American Institute of Physics, 2011), 157–71.

64. Camille Flammarion quoted in Glaisher, *Travels in the Air*, 148.

65. Dyson, *Tone of Our Times*, 2.

66. Parikka, *Geology of Media*.

67. Sasha Engelmann, "More-Than-Human Affinitive Listening," *Dialogues in Human Geography* 5, no. 1 (2015): 76–79.

68. Gallagher, Kanngieser, and Prior, "Listening Geographies."

69. Jol Thomson and Sasha Engelmann, "Intra-Acting with the IceCube Neutrino Observatory; Or How the Technosphere May Come to Matter," *Anthropocene Review* 4, no. 2 (2017): 81–91. Jol Thomson and Sasha Engelmann, "TIANHE ≋ 天河: Parables of the Celestial River," in *Aerocene Newspaper*, a publication accompanying Tomás Saraceno's *Aerocene* installation at the Grand Palais, Paris, December 2015. As Thomson and Engelmann demonstrate, the detection of neutrinos is an important example of this kind of terrestrial sensing. See also Y. Suzuki and K. Inoue, "Kamioka Underground Observatories," *European Physical Journal Plus* 127 (2012): 111–19.

70. For an account of this kind of field recording, see Michael Gallagher, "Field Recording and the Sounding of Spaces," *Environment and Planning D: Society and Space* 33, no. 3 (2015): 560–76.

71. John Durham Peters, *The Marvelous Clouds: Toward a Philosophy of Elemental Media* (Chicago, IL: University of Chicago Press, 2015).

1. Michel Serres, *The Five Senses: A Philosophy of Mingled Bodies*, trans. Margaret Sankey and Peter Cowley (London: Continuum, 2008), 302.

2. Michel Serres, *Biogea*, trans. Randolph Burks (Minneapolis, MN: Univocal Publishing, 2012), 74.

3. Serres, *Five Senses*, 146.

4. Serres, *Five Senses*, 146.

5. Luce Irigaray, *An Ethics of Sexual Difference*, trans. Carolyn Burke and Gillian C. Gill (London: Continuum, 2005), 7.

6. On the idea of suspension, see Timothy Choy and Jerry Zee, "Condition: Suspension," *Cultural Anthropology* 30, no. 2 (2015): 211.

7. Hugh Pearman, *Airports: A Century of Architecture* (London: Lawrence King, 2004).

8. Charles Coulston Gillespie, *Science and Polity in France: The End of the Old Regime* (Princeton, NJ: Princeton University Press, 2004).

9. Clifford Truesdell, "Jean-Baptiste Marie Charles Meusnier de la Place (1754–1793): An Historical Note," *Meccanica* 31 (1996): 610.

10. Vilém Flusser, *The Shape of Things: A Philosophy of Design*, trans. Anthony Mathews (London: Reaktion Books, 1999), 56.

11. Like many of these structures, Hangar Y has had an interesting afterlife. In the early 1960s the artist Marc Chagall was commissioned by the then French minister of culture, André Malraux, to paint the ceiling panels as part of the refurbishment of the opera house. Malraux had been impressed with Chagall's work for Maurice Ravel's ballet *Daphnis et Chloé*. For Chagall, the work was not just an aesthetic challenge. It also posed logistical difficulties, not least because the work would cover a surface area of 560 square meters. Chagall's work spaces were simply too small to accommodate the panels upon which he was working. Chagall needed a much bigger space. The space he found was Hangar Y, near Paris, which had been constructed in 1879 by Renard and Krebs to house their dirigible airship. Chagall produced the panels in the airship shed and, once completed, the panels were transported to the opera house. Walter Erben, *Marc Chagall* (New York: F. A. Praeger, 1966).

12. Peter Löscher, "MoU Signing," Siemens press release, June 17, 2013, http://www.siemens.com/press/pool/de/feature/2013/corporate/2013–06-airshow/speech-mou-signing-e.pdf, last accessed December 4, 2016.

13. Arthur A. Stuart, "A Nine-Acre Nest for Dirigibles," *Popular Science Monthly* (September 1929): 20.

14. Stuart, "Nine-Acre Nest," 21.

15. Reyner Banham, "Monumental Wind-Bags," in *The Inflatable Moment: Pneumatics and Protest in '68*, by Marc Dessauce (New York: Princeton Architectural Press, 1999), 33.

16. Eugène Freyssinet, "On the Sublime," *Architectural Research Quarterly* 5 (2001): 252.

17. Freyssinet, "On the Sublime," 253.

18. James Glaisher, ed., *Travels in the Air*, 2nd ed. (London: Richard Bentley and Son, 1871), viii.

19. Glaisher, *Travels in the Air*, viii.

20. "WAAFS Involved at RAF Chigwell in Making and Servicing Barrage Balloons," Barrage Balloon Reunion Club, http://www.bbrclub.org/WAAFS%20 involved%20at%20RAF%20Chigwell%20in%20making%20and%20servicing%20Bar- rage%20Balloons.htm, last accessed May 12, 2014.

21. In *Coming Down* (1941), the artist Robert Sargent Austin sketched a WAAF balloon-worker releasing the gas from a balloon as it deflates. In *In for Repairs* (1941), by Dame Laura Knight, nine women are involved in different aspects of the repair of a barrage balloon envelope.

22. Laura Levin, *Performing Ground: Space, Camouflage, and the Art of Blending In* (Manchester, UK: Palgrave Macmillan, 2014).

23. Interview with Hilda Mudd, Barrage Balloon Reunion Club, http://www .bbrclub.org/Hilda%20Mudd.htm, last accessed May 19, 2014.

24. In the second decade of the twentieth century, Knight became especially interested in the relation between painting and dance through the rhythmic move- ment of lines. Barbara Morden, *Laura Knight: A Life* (Pembroke, UK: McNidder and Grace, 2014); Laura Knight, *The Magic of a Line* (London: Kimber, 1965).

25. Knight, *Magic*, 203.

26. Tessa Stone, "Creating a (Gendered?) Military Identity: The Women's Auxil- iary Air Force in Great Britain in the Second World War," *Women's History Review* 8, no. 4 (1999): 611.

27. For a discussion of how working with balloons in a wartime context com- plicated some of the social stratification of bodies though race, see the account by Linda Hervieux of black US serviceman working with barrage balloons on D-Day in *Forgotten: The Untold Story of D-Day's Black Heroes* (New York: Harper, 2016).

28. William Forsythe, *Suspense: Exhibition Catalogue* (Zurich: JRP/Ringier Kun- stverlag, 2008), 5.

29. Henri Lachambre and Alexis Machuron, *Andrée's Balloon Expedition in Search of the North Pole* (New York: Frederick Stokes, 1898), 263–64.

30. This includes, for instance, the kind of aerial choreography required for the rigging of the cables in large suspension bridges. See David McCullough, *The Great Bridge: The Epic Story of the Building of the Brooklyn Bridge* (New York: Simon and Schuster, 1972), 380.

31. Peta Tait, *Circus Bodies: Cultural Identities in Aerial Performance* (London: Routledge, 2005).

32. Steven Connor, "Man Is a Rope," unpublished manuscript, 2008, http://www .stevenconnor.com/rope/, no pagination, last accessed May 13, 2014.

33. *The Balloonatic*, directed by Edward F. Cline and Buster Keaton (Los Angeles, CA: Buster Keaton Productions, 1923).

34. "A Big Echo's Little Pal," *Life Magazine*, June 21, 1963, 26.

35. "A Big Echo's Little Pal," 26.

36. "12 Miles High," 1933, Pathé Archive, http://www.britishpathe.com/video/12 -miles-high/query/Soviet, accessed May 12, 2014. Rex Hall and David Shayler, *The Rocket Men: Vostok and Voskhod, the First Soviet Manned Spaceflight* (Chichester, UK: Springer Praxis, 2001), 11.

37. Vida T. Johnson and Graham Petrie, *The Films of Andrei Tarkovksy: A Visual Fugue* (Bloomington: Indiana University Press, 2004), 219.

38. As such, *In Orbit* also points us to another sense in which we might understand what Serres calls the soft: the softness of messages. As Serres puts it, "Modernity is soft, and the message is soft, software is not hard. The modern object is not the steam engine, it is the computer, the manipulator of messages." Michel Serres, George Diez, and Christopher Roth, "The Neolithic Age Is Over," *032c* 20 (2010/2011), http://032c.com/2011/the-neolithic-age-is-over/, last accessed May 14, 2014.

39. Donald Barthelme, *Sixty Stories* (New York: Penguin, 2003), 47.

40. In an earlier work, *Poetic Cosmos of the Breath* (2007), Saraceno inflated a thin transparent polyester envelope in the dawn sunlight until it became a self-supporting structure around which visitors could walk. From the outside, billowing in the wind, the structure looked like a partially deflated balloon on the ground. "Poetic Cosmos of the Breath, Tomas Saraceno, 2007 (The Arts Catalyst)," YouTube, http://www.youtube.com/watch?v=qHOsO-IYpbw, last accessed May 5, 2014. The effect was monumental but also evanescent, a kind of gossamer-tented atmosphere of temporary occupation. In a more recent work, *On Space Time Foam* (2012), Saraceno played in a different way with inflation, pneumaticity, and the dynamic relation between bodies, movement, and surfaces. Three transparent "membranes" were stretched at different levels above a gallery at the Hangar Bicocca in Milan, formerly the site of a manufacturing facility for train components. Visitors could move across these different membranes, generating shapes and forms of change across and between bodies, surfaces, and volumes. The installation was a leaky, dynamic set of volumes that worked through the dynamic interplay among the material fabric, moving bodies, and the pressure and volume of the air inside and outside the building.

41. "On Space Time Foam / Tomás Saraceno," YouTube, http://www.youtube .com/watch?v=MIPLjOGvhtM, last accessed May 5, 2014.

42. For an overview of Saraceno's work, see Tomás Saraceno, *Cloud Cities*, ed. Marion Ackerman, Daniel Birnbaum, Udo Kittelmann, and Hans Ulrich Obrist (Berlin: Distanz Verlag, 2011).

43. Sasha Engelmann, "More-Than-Human Affinitive Listening." *Dialogues in Human Geography* 5, no. 1 (2015): 76–79; Sasha Engelmann, *The Cosmological Aesthetics of Tomás Saraceno's Atmospheric Experiments*. DPhil Thesis, Oxford University, 2017.

44. An emphasis on elasticity and tension is evident in other works by Saraceno, most obviously, perhaps, his *Galaxies Forming along Filaments, Like Droplets along the Strands of a Spider's Web* (2008).

45. Irigaray, *Ethics of Sexual Difference*, 7.

46. Forsythe, *Suspense*, 7.

1. *The Prisoner*, which aired from September 1967 until February 1968, was created by Patrick McGoohan and George Markstein.

2. Peter Adey, "Security Atmospheres or the Crystallization of Worlds," *Environment and Planning D: Society and Space* 32, no. 5 (2014): 834–51.

3. Caren Kaplan, "Mobility and War: The Cosmic View of US Air Power," *Environment and Planning A* 38, no. 2 (2006): 395–407; Caren Kaplan, "'A Rare and Chilling View': Aerial Photography as Biopower in the Visual Culture of '9/11,'" *Reconstruction* 11, no. 2 (2011); Derek Gregory, "From a View to a Kill: Drones and Late Modern War," *Theory, Culture, and Society* 28, nos. 7–8 (2011): 188–215. Peter Adey, Mark Whitehead, and Alison J. Williams, "Introduction: Air-Target Distance, Reach and the Politics of Verticality," *Theory, Culture, and Society* 28, nos. 7–8 (2011): 173–87; Paul Virilio, *The Vision Machine* (Bloomington: Indiana University Press, 1994).

4. Lisa Parks and Caren Kaplan, eds. *Life in the Age of Drone Warfare* (Durham, NC: Duke University Press, 2017).

5. Ian Shaw, *Predator Empire: Drone Warfare and Full Spectrum Dominance* (Minneapolis: University of Minnesota Press, 2016).

6. "JLENS," Raytheon, http://www.raytheon.com/capabilities/products/jlens/, last accessed December 6, 2016.

7. David Willman, "How a $2.7 Billion Air Defense System Became a 'Zombie' Program," *Los Angeles Times*, September 24, 2015, http://graphics.latimes.com /missile-defense-jlens/, last accessed November 11, 2015.

8. Andreas Peterson, Craig Timberg, and Christian Davenport, "The Military Lost Control of a Giant Unmanned Surveillance Blimp," *Washington Post*, October 28, 2015, https://www.washingtonpost.com/news/the-switch/wp/2015/10/28 /the-army-lost-control-of-a-giant-unmanned-surveillance-blimp/, last accessed November 10, 2015.

9. To be sure, of course, there have been numerous efforts to use dirigible airships as platforms for the delivery of military ordinance and for the generation of affects, not least the use of German zeppelins over London during World War I. See Peter Adey, *Aerial Life: Mobilities, Spaces, Affects* (London: Wiley-Blackwell, 2010). These dirigibles have been understood as part of the genealogy of the drone, albeit perhaps too neatly, and too easily. Derek Gregory, "Moving Targets and Violent Geographies," in *Spaces of Danger: Culture and Power in the Everyday*, ed. Heather Merrill and Lisa Hoffman (Athens: University of Georgia Press, 2015), 256–97. The possibility that Goodyear blimps could be put to use as weapons of terror was the scenario around which the 1977 film *Black Sunday* was based. In the film, Bruce Dern plays a disgruntled Goodyear blimp pilot in league with a Palestinian terrorist. Together they plan to detonate a bomb above the stadium during a Super Bowl game in Miami, a plot that is inevitably foiled.

10. Francesco Lana de Terzi, *The Aerial Ship* (London: Aeronautical Society of Great Britain, 1910), 26–27.

11. Charles Ziegler, "Weapons Development in Context: The Case of the World War I Balloon Bomber," *Technology and Culture* 35, no. 4 (1994): 750–67.

12. Ziegler, "Weapons Development."

13. Ziegler, "Weapons Development," 759.

14. As Ziegler writes, plans for the preemptive use of balloons implied "a conviction that breaking the formerly accepted "rules of war" was characteristic of the enemy. It is a version of the logic of reprisal in which it is permissible to break the rules if the enemy breaks them first (or can be expected to)." Ziegler, "Weapons Development," 765.

15. Ziegler, "Weapons Development," 765.

16. Peter Sloterdijk, "Airquakes," *Environment and Planning D: Society and Space*, 27 (2009): 41–57.

17. Cornelius Conley, "The Great Japanese Balloon Offensive," *Air University Review*, January–February 1968, http://www.airpower.maxwell.af.mil/airchronicles /aureview/1968/jan-feb/conley.html, last accessed May 26, 2014.

18. "Saw Wife and Five Children Killed by Jap Balloon," *Seattle Times*, June 1, 1945.

19. A twenty-minute training film for navy personnel detailing the operation of the balloons ended with the following message: "Any balloons approaching the United States from outside its borders can be enemy attacks against the nation. Information that the balloons have reached this country and particularly what section they have reached is information of value to the enemy. Please do not aid the enemy by publishing or broadcasting or discussing such information without appropriate authority." "Japanese Paper Balloon," United States Navy Training Film, 1945, http://www.youtube.com/watch?v=xmW-SgNqTRc, last accessed November 11, 2015.

20. Even if reporting restrictions meant that the Japanese fire balloons never modified the affective life, or morale, of the population to any significant extent, they did contribute to the circulation of rumors of other kinds of atmospheric things. Part of the lore surrounding these balloons and others is that they were sometimes mistaken for unidentified flying objects, spectral forms in the air above the United States.

21. Alan Riches, "Balloons: What Have They Ever Done for Us? The Contribution of the Balloon to the History of Air Power," *Royal Air Force Historical Society Journal* 27 (2002): 24–42.

22. "Trailing Wires Cut Power Lines Serving Railways," *Independent* (St. Petersburg, Florida), Wednesday September 18, 1940, 9, http://news .google.com/newspapers?nid=950&dat=19400918&id=4-FPAAAAIBAJ&sjid =rFQDAAAAIBAJ&pg=4922,4387654, last accessed May 26, 2014.

23. For a detailed discussion of this aspect of the project, see Raoul E. Drapeau, "Operation Outward: Britain's World War II Offensive Balloons," *IEEE Power and Energy Magazine*, September/October 2011, 94–105.

24. Drapeau, "Operation Outward."

25. During the Siege of Paris in 1870 balloons were used to carry both people and post out of the city. On these flights those in the balloon often took the opportunity to drop leaflets onto the Prussian troops below. While a limited number of propaganda leaflets were also dropped from balloons during World War I, it was not until World War II that experiments with this tactic were undertaken on a much larger scale as part of British efforts to distribute propaganda across Germany. See Lee Richards, "The 'M'-Balloon Unit: British Balloon Distribution of Aerial Propaganda during WWII," http://www.psywar.co.uk/psywar/reproductions/mballoon.pdf, last accessed May 26, 2014. On more contemporary uses of leaflets and targeting morale, see Ben Anderson, "Morale and the Affective Geographies of the 'War on Terror,'" *Cultural Geographies* 17, no. 2 (2010): 219–36; and Alasdair Pinkerton, Stephen Young, and Klaus Dodds, "Postcards from Heaven: Critical Geographies of the Cold War Military-Industrial-Academic Complex," *Antipode* 43, no. 3 (2011): 820–44.

26. Richards, "'M'-Balloon Unit," 10.

27. Sir Campbell Stuart quoted in Richards, "'M'-Balloon Unit," 5.

28. "Balloon Barrage Grows: 6,300,000 Messages of Hope, Friendship sent to Czechs," *New York Times*, August 19, 1951, 43.

29. Drew Pearson quoted in Richard H. Cummings, *Radio Free Europe's "Crusade for Freedom": Rallying Americans behind Cold War Broadcasting, 1950–1960* (Jefferson, NC: McFarland, 2010), 48.

30. For a discussion of the wider relations between science and the Cold War, see Naomi Oreskes and John Krige, eds, *Science and Technology in the Global Cold War* (Cambridge, MA: MIT Press, 2014); Trevor Barnes and Matthew Farish, "Between Regions: Science, Militarism, and American Geography from World War to Cold War." *Annals of the Association of American Geographers* 96, no. 4 (2006): 807–26.

31. Pinkerton, Young, and Dodds, "Postcards."

32. Cummings, *Radio Free Europe*, 15.

33. Quoted in Richard H. Cummings, *Cold War Radio: The Dangerous History of American Broadcasting in Europe, 1950–1989* (Jefferson, NC: McFarland, 2009), 10. Allan Needell, "Truth Is Our Weapon": Project Troy, Political Warfare, and Government-Academic Relations in the National Security State," *Diplomatic History* 17, no. 3 (1993): 399–420.

34. Memo from C. Jackson to FEC staff about Operation WINDS OF FREEDOM, August 14, 1951, Folder 151.7, "Balloons Czechoslovakia 1951–1953," Box 151, Radio Free Europe / Radio Liberty Corporate Records, Hoover Institution Archives, 1.

35. "Operation Focus: September 15–December 31 Report," Folder 152.9, "Balloons Hungary Operation Focus Review," Box 152, Radio Free Europe/Radio Liberty Corporate Records, Hoover Institution Archives, 17.

36. Memo from C. Jackson, August 15, 1951, Folder 151.7, "Balloons Czechoslovakia 1951–1953," Box 151, Radio Free Europe/Radio Liberty Corporate Records, Hoover Institution Archives, 3.

37. Operation Veto report, September 17, 1954, Folder 152.1, "Balloons Czecho-slovakia Operation Veto Effectiveness Studies," Box 152, Radio Free Europe/Radio Liberty Corporate Records, Hoover Institution Archives, 20.

38. Operation Veto report, 3.

39. Operation Veto report, 23.

40. Pinkerton, Young, and Dodds, "Postcards."

41. Operation Veto report, 11.

42. Operation Veto report, 11.

43. Operation Veto report, 11.

44. Cummings, *Radio Free Europe*, 6–7.

45. Cummings, *Radio Free Europe*, 6–7.

46. Cummings, *Radio Free Europe*, 6–7.

47. Cummings, *Radio Free Europe*, 34.

48. Cummings, *Radio Free Europe*, 64.

49. Cummings, *Radio Free Europe*, 64.

50. Operation Veto report, 44.

51. "Evaluation Report: Internal Distribution of Operation Veto Leaflets and Concepts, Research Analysis Department," Folder 151.9, "Balloons Czechoslovakia Operation Veto 1954–1955," Box 151, Radio Free Europe / Radio Liberty Corporate Records, Hoover Institution Archives, 1.

52. Memo from Dunning to Rafael/Hennon/Egan, February 20, 1956, Folder 151.10, "Balloons Czechoslovakia Operation Veto Effectiveness," Box 151, Radio Free Europe/Radio Liberty Corporate Records, Hoover Institution Archives, 1.

53. Operation Veto report, 39.

54. There is another story to be told of the launch of these balloons, one beyond the scope of my concerns here. This is a story of how the release of these balloons generated all kinds of diplomatic and geopolitical tensions in Cold War Europe.

55. Merton E. Davies and William R. Harris, "RAND's Role in the Evolution of Balloon and Satellite Observation Systems and Related US Space Technology," (Santa Monica, CA: RAND Corporation, 1988), 20.

56. Gregory Pedlow and Donald Welzenbach, *The Central Intelligence Agency and Overhead Reconnaissance: The U-2 and OXCART Programs, 1954–1974* (Washington, DC: Central Intelligence Agency, 1992), 18. This document was declassified in 2013.

57. Pedlow and Welzenbach, *Central Intelligence Agency*, 18.

58. Pedlow and Welzenbach, *Central Intelligence Agency*, 18.

59. Davies and Harris, "RAND's Role," 39. Nevertheless, some important insights were generated by the project. The photographs were the best images that had yet been obtained of the Soviet Union. And the experience of tracking the balloons also generated understandings of the patterns of high-altitude winds that were to prove useful in the planning of U-2 operations in the following years. More generally, however the program of the balloon experiments had another payoff. These experiments "encouraged the design of lightweight, durable subsystems" and

provided a "technology bridge" from aircraft-mounted to satellite-borne reconnaissance systems.

60. Gilles Deleuze and Félix Guattari, *A Thousand Plateaus*, trans. Brian Massumi (London: Athlone, 1987).

61. Deleuze and Guattari, *Thousand Plateaus*, 291.

62. On September 12, 1995, Alan Fraenkel and John-Stuart Jervis were killed when a Belarus attack helicopter shot down the balloon in which they were traveling. They had been participants in the Gordon Bennett Cup, first held in 1906. Despite long periods during which its staging has been suspended, the race for the Gordon Bennett Cup is still held every year, and continues to be organized according to the same basic rules as the 1906 race: the winners are those who can travel farthest with the help of the winds in a gas-filled free balloon. In the early decades of the twentieth century, the race could be and was understood as a kind of experiment in meteorology. For the winner of the race in 1913, R. H. Upson, balloon racing was a "game of practical meteorology" that might help the broader development of aeronautics. Maps of the tracks and the journeys of the balloons were of interest to meteorologists, and were sometimes published in meteorology journals. If it was an exercise that could be understood as part of the rendering explicit of the meteorological atmosphere, the Gordon Bennett Cup was also linked closely with the generation of affective atmospheres through systems of news media distribution. The race was originally sponsored by James Gordon Bennett, Jr., son of the founder and owner of the *New York Herald*, James Gordon Bennett, Sr. Like other newspaper proprietors, Bennett and his son understood the value of event-based sensationalism, and knew that accounts of derring-do in the air made for good stories: to this end, for instance, Bennett Sr. had sponsored Henry Morton Stanley's expedition to Africa to find missionary David Livingstone. R. H. Upson, quoted in M. A. Giblet, "The International Free-Balloon Race for the Gordon Bennett Cup," *Quarterly Journal of the Royal Meteorological Society* 50 (1924): 260–67.

63. Cristina Corbin, "Bible Drop: Christian Group Takes to the Sky to Sneak Bibles into North Korea," *Fox News*, http://www.foxnews.com/world/2013/11/08/bible-drop-christian-group-takes-to-sky-to-sneak-gospel-into-north-korea/?intcmp=latestnews, last accessed Thursday, January 9, 2014.

64. "Park Sang-hak on Receiving the Václav Havel Prize for Creative Dissent," *News Focus International*, May 7, 2013, http://newfocusintl.com/park-sang-hak-vaclav-havel-prize-for-creative-dissent/, last accessed January 9, 2014.

65. "Park Sang-Hak."

66. Mark Donald, "Balloon-Borne Messages to North Korea Have Detractors on Both Sides of the Border, *New York Times*, April 26, 2011, http://www.nytimes.com/2011/04/27/world/asia/27iht-korea.html?pagewanted=all&_r=0Mark, last accessed January 9, 2014.

67. David Long, "CBP's Eyes in the Sky: CBP's Tethered Aerostats Keep Watch for Trouble from 10,000ft," *U.S. Customs and Border Protection*, September 13, 2016, https://www.cbp.gov/frontline/frontline-november-aerostats, last accessed October 16, 2016.

68. Yaakov Lappin, "A Common Sight during the Gaza War, IDF's Reliance on Aerostat Balloons Is Up," *Jerusalem Post*, June 10, 2014, http://www.jpost.com /Israel-News/New-Tech/IDFs-reliance-on-aerostats-is-up-378127, last accessed, October 16, 2016.

69. Amitai Ziv, "24/7 Big Brother: Gaza—The Technical Means That Observe the Gaza Strip 24 Hours a Day," *Marker*, August 1, 2014, http://www.rt.co.il/themarker, last accessed October 16, 2016.

70. Khalil Ashraf, "Palestinians Mark a Catastrophe," *Los Angeles Times*, May 16, 2008, http://articles.latimes.com/2008/may/16/world/fg-palestinians16, last accessed October 18, 2016.

71. Sam Kishawi, "Lifting Spirits: Chicago's Balloon Release for Gaza," https:// smpalestine.com/2011/12/31/lifting-spirits-chicagos-balloon-release-for-gaza/, last accessed June 1, 2017.

72. Patricia Pisters, *The Neuro-Image: A Deleuzian Film-Philosophy of Digital Screen Culture* (Stanford, CA: Stanford University Press, 2012), 255.

73. Alfredo Jaar, *The Cloud*, 2000, Valle del Matador, Tijuana, US-Mexico border, public intervention. For a discussion of *The Cloud* in the context of the US-Mexico border as a site of performative intervention, see Ila Nicole Sheren, *Portable Borders: Performance Art and Politics on the Frontera since 1984* (Austin: University of Texas Press, 2015); and Elena Stromberg, "'I Will Not Act before Understanding: Context Is Everything': The Work of Alfredo Jaar," *Los Angeles Review of Books*, December 8, 2013.

74. Clearly, such events of balloon release are planned as political acts, intended to foreground matters of controversy or injustice via the generation of atmospheric things. It will be interesting to observe how long it takes for the environmental implications of such acts to make them ethically unacceptable.

75. Sheren, *Portable Borders*, 106.

76. Sheren, *Portable Borders*, 109.

CHAPTER 9. ELEMENTS

1. Peter Adey, "Air's Affinities: Geopolitics, Chemical Affect, and the Force of the Elemental," *Dialogues in Human Geography* 5, no. 1 (2015): 54–75. Indeed, a revival of scholarly engagement with the elemental is part of a more general interest within the social sciences and humanities with the turbulent spatiotemporality of materiality. See also Gabrielle Hecht, *Nuclear Being: Africans and the Global Uranium Trade* (Cambridge, MA: MIT Press, 2012); Jeffrey Cohen and Lowell Duckert, eds., *Elemental Ecocriticism: Thinking with Earth, Air, Water, Fire* (Minneapolis: University of Minnesota Press, 2015); Rachel Squire, "Rock, Water, Air and Fire: Foreground- ing the Elements in the Gibraltar-Spain Dispute," *Environment and Planning D: Society and Space* 34, no. 3 (2016): 545–63.

2. *The Big Bounce*, directed by Jerry Fairbanks United States: Jerry Fairbanks Productions, 1960.

3. Adey, "Air's Affinities"; Hecht, *Nuclear Being*; Cohen and Duckert, *Elemental Ecocriticism*.

4. David Macauley, *Elemental Philosophy: Earth, Air, Fire, and Water as Environmental Ideas* (Albany, NY: SUNY Press, 2010).

5. Tim Ingold, "Footprints Through the Weather-world: Walking, Breathing, Knowing," *Journal of the Royal Anthropological Institute* 16, no. 1 (2010): 121–39.

6. Roger Barry and Richard Chorley, *Atmosphere, Weather, and Climate*, 9th ed. (London: Routledge, 2009).

7. The existence of this layer was noted also by German meteorologist Richard Assmann. F. K. Hare, "The Stratosphere," *Geographical Review* 52, no. 1 (1962): 525–47; R. M. Goody, *The Physics of the Stratosphere* (Cambridge: Cambridge University Press, 1958).

8. De Bort, cited in Andrew Watt, "The Exploration of the Upper Air," *Bulletin of the American Geographical Society* 42, no. 1 (2010): 37–51, 49.

9. Elsewhere I explore elements of Malcolm D. Ross's ascents into the stratosphere, and the particular form of atmospheric address they suggest. Derek McCormack, "Stratospheric Envelopes for an Atmospheric Mode of Address," *GeoHumanities*, forthcoming, http://dx.doi.org/10.1080/2373566X.2017.1301214. Jeannette Piccard's story is especially interesting, not least because it reveals the gendered dimension of scientific ballooning during this period. See Sheryl Hill, "'Until I Have Won': Vestiges of Coverture and the Invisibility of Women in the Twentieth Century: A Biography of Jeannette Ridlon Piccard" (PhD thesis, Ohio University, 2009).

10. Brad Stone, "Inside Google's Secret Lab," *Bloomberg Business Week*, May 22, 2013.

11. "Project Loon: Google Balloon That Beams Down Internet Reaches Sri Lanka," *Guardian*, February 16, 2016, http://www.theguardian.com/technology /2016/feb/16/project-loon-google-balloon-that-beams-down-internet-reaches-sri -lanka, last accessed March 9, 2016; and "Google Installing Hundreds of Internet Enabled Balloons in Indonesia," *Guardian*, October 28, 2015, http://www .theguardian.com/technology/2015/oct/28/google-installing-20000-internet -enabled-balloons-in-indonesia, last accessed March 8, 2016.

12. Flightradar24.com, accessed September 21, 2017, at GMT 14:23. https://www .flightradar24.com/HBAL247/eeec1a8.

13. "Astro Teller: The Unexpected Benefit of Celebrating Failure," www.ted.com /talks/astro_teller_the_unexpected_benefit_of_celebrating_failure#t-14599, last accessed September 9, 2016.

14. Caren Kaplan, "Mobility and War: The Cosmic View of US Air Power," *Environment and Planning A* 38, no. 2 (2006): 395–407; Derek Gregory, "From a View to a Kill: Drones and Late Modern War," *Theory, Culture, and Society* 28, nos. 7–8 (2011): 188–215.

15. Stuart Banner, *Who Owns the Sky? The Struggle to Control Airspace from the Wright Brothers On* (Cambridge, MA: Harvard University Press, 2008); Stuart Elden, "Secure the Volume: Vertical Geopolitics and the Depth of Power," *Political Geography* 34, no. 1 (2013): 35–51.

16. Andreas Philippopoulos-Mihalopoulos, "Withdrawing from Atmosphere: An Ontology of Air Partitioning and Affective Engineering." *Environment and Planning D: Society and Space* 34, no. 1 (2016): 150–67.

17. Will Butler, "Can We Trust Google with the Stratosphere?" *Atlantic* (2013), www.theatlantic.com/technology/archive/2013/08/can-we-trust-google-with-the -stratosphere/278797/, last accessed March 7, 2016; George John, "Welcome to the Space Jam: How United States Regulators Should Govern Google and Facebook's New Internet-Providing High-Altitude Platforms," *American University Business Law Review* 471 (2015): 471–504.

18. "Delivering Connectivity," YouTube, https://www.youtube.com/watch?v =BECoG2HbuiE, last accessed September 9, 2016.

19. Casper Bruun Jenson and Atsuro Morita, "Infrastructures as Ontological Experiments," *Ethnos* 82, no. 4 (2017): 615–26.

20. Heyward Ehrlich, "Poe in Cyberspace: Balloons! Drones!! The Global Internet!!!" *Edgar Allan Poe Review* 16, no. 2 (2015): 242–46.

21. Ben Popper, "Inside Project Loon: Google's Internet in the Sky Is Almost Open for Business," *Verge*, March 2, 2015, http://www.theverge.com/2015/3/2 /8129543/google-x-internet-balloon-project-loon-interview, last accessed September 9, 2016.

22. Popper, "Inside Project Loon."

23. W. Butler, "Can We Trust Google?"; see also Adam Sniderman, Mireille Broucke, and Gabriele D'Eleuterio, "Formation Control of Balloons: A Block Circulant Approach," in *Proceedings of the American Control Conference* (Chicago, IEEE, 2015), 1463–68.

24. "Delivering Connectivity."

25. "Project Loon: The Technology," YouTube, https://www.youtube.com/watch ?v=mcw6j-QWGMo/, accessed November 2, 2016.

26. "Project Loon: The Technology."

27. Mark Hansen, *Feed-Forward: On the Future of Twenty-First-Century Media* (Chicago, IL: University of Chicago Press, 2015), 183.

28. Astro Teller, "How Project Loon's Smart Software Learned to Sail the Winds," Google X Blog, February 16, 2017, https://blog.x.company/how-project -loons-smart-software-learned-to-sail-the-winds-ec904e6d08c, last accessed June 1, 2017.

29. Cade Metz, "Machine Learning Invades the Real World on Internet Balloons," *Wired Magazine*, February 2, 2017, https://www.wired.com/2017/02/machine -learning-drifting-real-world-internet-balloons/, last accessed June 1, 2017.

30. Cymene Howe and Dominic Boyer, "Aeolian Politics," *Distinktion: Scandinavian Journal of Social Theory* 16, no. 1 (2015): 31–48, 44.

31. Hansen, *Feed-Forward*.

32. Alexander Galloway and Eugene Thacker, *The Exploit: A Theory of Networks* (University of Minnesota Press: Minneapolis, 2007), 157.

33. Stephen Groening, "Towards a Meteorology of the Media," *Transformations* 25 (2014): n.p.

34. Michel Serres, *The Five Senses: A Philosophy of Mingled Bodies*, trans. Margaret Sankey and Peter Cowley (London: Athlone Press, 2008); John Durham Peters, *The Marvelous Clouds: A Philosophy of Elemental Media* (Chicago, IL: University of Chicago Press, 2015).

35. Levi Bryant, *Onto-Cartography: An Ontology of Machines and Media* (Edinburgh, Scotland: Edinburgh University Press, 2014), 193.

36. Bryant, *Onto-Cartography*, 3.

37. Peters, *Marvelous Clouds*, 105.

38. Tom Roberts, "From Things to Events: Whitehead and the Materiality of Process," *Environment and Planning D: Society and Space* 32, no. 6 (2014): 968–83.

39. This elemental commons overlaps in some ways with the idea, albeit different, of an atmospheric and perhaps ambient commons. Malcolm McCullough, *Ambient Commons: Attention in the Age of Embodied Information* (Cambridge, MA: MIT Press, 2013).

40. Carolin Weidemann, "Between Swarm, Network, and Multitude: Anonymous and the Infrastructures of the Commons," *Distinktion* 15, no. 3 (2014): 309–26.

41. James Ash, *The Interface Envelope: Gaming, Technology, Power* (London: Bloomsbury, 2015).

42. Nick Shapiro, "Alter-Engineered World," in *Aerocene Newspaper*, publication accompanying Tomás Saraceno's *Aerocene* installation at the Grand Palais, Paris, December 2015. It is worth noting that in July 2017 one of the most important Loon patents was reportedly cancelled after a lawsuit filed by a small US-based company called Space Data. This company claimed that they had come up with the idea of changing a balloon's direction by altering its altitude. See Monica Alleven, "Space Data Strikes Win in Getting Key Project Loon Patent Canceled," Fierce Wireless, July 10, 2017. http://www.fiercewireless.com/wireless/space-data-strikes-win-getting-key-project-loon-patent-canceled. Last accessed September 21, 2017.

43. Lauren Berlant, "The Commons: Infrastructures for Troubling Times," *Environment and Planning D: Society and Space* 34, no. 3 (2016): 394.

44. Berlant, "Commons," 394.

45. AbdouMaliq Simone, "Passing Things Along: (In)completing Infrastructure," *New Diversities* 17, no. 2 (2015): 151.

46. See the campaign of the UK-based Marine Conservation Society against the release of balloons and sky lanterns.

47. Sasha Engelmann, "Toward a Poetics of Air: Sequencing and Surfacing Breath," *Transactions of the Institute of British Geographers* 40, no. 3 (2015): 430–44.

48. Rachel Armstrong, *Vibrant Architecture: Matter as a CoDesigner of Living Structures* (Berlin: De Gruyter Open, 2015), 173.

49. Etienne Turpin, "Aerosolar Infrastructure: Polities above and beyond Territory," in *Becoming Aerosolar*, by Tomás Saraceno, 169–83 (Vienna: 21er Haus, 2015).

50. Bruno Latour, "Some Experiments in Art and Politics," *E-Flux Journal*, March 2011, http://www.e-flux.com/journal/some-experiments-in-art-and-politics/; and Bruno Latour, "Network Theory / Networks, Societies, Spheres: Reflections

of an Actor-Network Theorist," *International Journal of Communication* 5 (2011): 796–810.

51. Sasha Engelmann and Derek McCormack, "Elemental Aesthetics: On Artistic Experiments with Solar Energy," *Annals of the Association of American Geographers* 108, no. 1 (2018); 241–59. See also Sasha Engelmann, "More-Than-Human Affinitive Listening," *Dialogues in Human Geography* 5, no. 1 (2015): 76–79. My comments in this present section draw upon ideas developed in a number of invited contributions to pieces about Saraceno's work.

52. On the concepts of "cloud cities" and "becoming aerosolar," see Tomás Saraceno, Sasha Engelmann, and Bronislaw Szerszynski, "Becoming Aerosolar: From Solar Sculptures to Cloud Cities," in *Art in the Anthropocene: Encounters among Aesthetics, Politics, Environments, and Epistemologies*, ed. Heather Davis and Etienne Turpin (London: Open Humanities Press, 2015), 57–62.

53. Gilles Deleuze and Félix Guattari, *A Thousand Plateaus: Capitalism and Schizophrenia*, trans. Brian Massumi (Minneapolis: University of Minnesota Press, 1987), 142.

54. Monck Mason, *Aeronautica; Or, Sketches Illustrative of the Theory and Practice of Aerostation: Comprising an Enlarged Account of the Late Aerial Expedition to Germany* (London: F. C. Westley, 1838).

55. Félix Guattari, *Chaosmosis: An Ethico-Aesthetic Paradigm*, trans. Paul Bains and Julian Pefanis (Sydney: Power, 1995).

56. For details of this collective experiment, see http://aerocene.org/, last accessed September 11, 2017.

57. Jill Bennett, *Practical Aesthetics: Events, Affects, and Art after 9/11* (London: I. B. Tauris, 2012), 3.

58. In doing so, *Museo Aero Solar* obviously uses a material—plastic—produced through other forms of energy. But it takes this material as an already produced artifact of the Anthropocene.

59. On how Saraceno's work exemplifies a kind of associative elementalism, see Engelmann, McCormack, and Szerszynski, "Becoming Aerosolar," 67–101. Another way to think of this is as a version of what, following Jordan Crandall, Mark Hansen terms "gatherings." Rather than organized spaces of collective experience, these gatherings are better understood as "continuously and incrementally renewed, distinctly horizontal and anticipatory compositions" that draw together bodies, materials, and elements in different and circumstance-specific envelopes of association. Hansen, *Feed-Forward*, 257; Jordan Crandall, "Summary of Gatherings," http://jordancrandall.com/main/+GATHERINGS/index.html, last accessed November 12, 2015.

60. Luce Irigaray, *An Ethics of Sexual Difference*, trans. Carolyn Burke and Gillian C. Gill (London: Continuum, 2005), 213.

61. Frances Dyson, *Sounding New Media: Immersion and Embodiment in the Arts and Culture* (Berkeley: University of California Press, 2009), 16–17.

62. Durham Peters, *Marvelous Clouds*, 380.

63. Durham Peters, *Marvelous Clouds*, 105.

64. Adey, "Air's Affinities."

65. Nigel Clark, *Inhuman Natures: Sociable Life on a Dynamic Planet* (London: Sage, 2011). See also Jussi Parikka, *A Geology of Media* (Minneapolis: University of Minnesota Press, 2015).

66. Andreas Philippopoulos-Mihalopoulos, *Spatial Justice: Body, Lawscape, Atmosphere* (London: Routledge, 2015).

67. Edward Wong, "Beijing Issues Red Alert over Air Pollution for the First Time," *New York Times*, December 7, 2015.

68. Timothy Choy, *Ecologies of Comparison: An Ethnography of Endangerment in Hong Kong* (Durham, NC: Duke University Press, 2011).

69. For an account of the micropolitical, see Thomas Jellis and Joe Gerlach, "Micropolitics and the Minor," *Environment and Planning D: Society and Space* 35, no. 4 (2017): 563–67.

70. Hansen, *Feed-Forward*, 197.

71. Dyson, *Sounding New Media*, 16.

72. Ian Shaw, *Predator Empire: Drone Warfare and Full Spectrum Dominance* (Minneapolis: University of Minnesota Press, 2016).

73. Ray Bradbury, *The Martian Chronicles* (London: Harper Collins, 2008).

74. Irigaray, *Ethics of Sexual Difference*, 16; italics in source.

75. Irigaray, *Ethics of Sexual Difference*, 16.

76. Penelope Ingram, "From Goddess Spirituality to Irigaray's Angel: The Politics of the Divine," *Feminist Review* 66 (autumn 2000): 46–72; and Lenart Skof and Emily Holmes, eds., *Breathing with Luce Irigaray* (London: Bloomsbury, 2013).

77. Michel Serres, *Angels: A Modern Myth*, trans. Francis Cowper (Paris: Flammarion, 1995), 9.

78. Serres, *Angels*, 8.

79. Serres, *Angels*, 8.

80. Serres, *Angels*, 25.

81. Michel Serres, *Biogea*, trans. Randolph Burks (Minneapolis, MN: Univocal Publishing, 2012), 74.

82. Durham Peters, *Marvelous Clouds*, 33.

Adey, Peter. *Aerial Life: Mobilities, Spaces, Affects*. London: Wiley-Blackwell, 2010.

———. "Air/Atmospheres of the Megacity." *Theory, Culture, and Society* 30, nos. 7–8 (2013): 291–308.

———. "Air's Affinities: Geopolitics, Chemical Affect, and the Force of the Elemental." *Dialogues in Human Geography* 5, no. 1 (2015): 54–75.

———. "Securing the Volume/Volumen: Comments on Stuart Elden's Plenary Paper 'Secure the Volume.'" *Political Geography* 34, no. 1 (2013): 52–54.

———. "Security Atmospheres or the Crystallization of Worlds." *Environment and Planning D: Society and Space* 32, no. 5 (2014): 834–51

Adey, Peter, Laure Brayer Laure, Damien Masson, Patrick Murphy, Paul Simpson, and Nicholas Tixier. "'Pour votre tranquilité': Ambiance, Atmosphere, and Surveillance." *Geoforum* 49 (2013): 299–309.

Adey, Peter, Mark Whitehead, and Alison J. Williams. "Introduction: Air-Target Distance, Reach and the Politics of Verticality." *Theory, Culture, and Society* 28, nos. 7–8 (2011): 173–87.

Ahmed, Sara. "Happy Objects." In *The Affect Theory Reader*, edited by Melissa Gregg and Gregory Seigworth, 29–51. Durham, NC: Duke University Press, 2010.

Allen, John. "Ambient Power: Berlin's Potsdamer Platz and the Seductive Logic of Public Spaces." *Urban Studies* 43, no. 2 (2006): 441–55.

Amad, Paula. "From God's-Eye to Camera-Eye: Aerial Photography's Post-Humanist and Neo-Humanist Visions of the World." *History of Photography* 36 no. 1 (2012): 66–86.

Anderson, Ben. "Affective Atmospheres." *Emotion, Space, and Society* 2 no. 2 (2009): 77–81.

———. *Encountering Affect: Capacities, Apparatuses, Conditions*. Farnham, UK: Ashgate, 2014.

———. "Morale and the Affective Geographies of the 'War on Terror.'" *Cultural Geographies* 17, no. 2 (2010): 219–36.

Anderson, Ben, and James Ash. "Atmospheric Methods." In *Nonrepresentational Methods: Re-Envisioning Research*, edited by Philip Vannini, 34–51. London: Routledge, 2015.

Anderson, Ben, and John Wylie. "On Geography and Materiality." *Environment and Planning A* 41, no. 2 (2009): 318–35.

Armstrong, Rachel. *Vibrant Architecture: Matter as a CoDesigner of Living Structures.* Berlin: De Gruyter Open, 2015.

Artaud, Antonin. "There Is No More Firmament." In *Collected Works.* Vol. 2. Translated by Victor Corti. London: Calder and Boyars, 1971.

Ash, James. *The Interface Envelope: Gaming, Technology, Power.* London: Bloomsbury, 2015.

———. "Rethinking Affective Atmospheres: Technology, Perturbation and Space-Times of the Non-Human." *Geoforum* 49, no. 1 (2013): 20–28.

———. "Technologies of Captivation Videogames and the Attunement of Affect." *Body and Society* 19, no. 1 (2013): 27–51.

Ashraf, Khalil. "Palestinians Mark a Catastrophe." *Los Angeles Times*, May 16, 2008. http://articles.latimes.com/2008/may/16/world/fg-palestinians16. Last accessed October 18, 2016.

Bachelard, Gaston. *Air and Dreams: An Essay on the Imagination of Movement.* Translated by Edith R. Farrell and C. Frederick Farrell. Dallas, TX: Dallas Institute Publications, 1988.

"Balloon Barrage Grows: 6,300,000 Messages of Hope, Friendship Sent to Czechs." *New York Times*, August 19, 1951, 43.

Banham, Reyner. "Monumental Wind-Bags." In *The Inflatable Moment: Pneumatics and Protest in '68*, edited by Marc Dessauce, 31–33. New York: Princeton Architectural Press, 1999.

Banner, Stuart. *Who Owns the Sky? The Struggle to Control Airspace from the Wright Brothers On.* Cambridge, MA: Harvard University Press, 2008.

Barber, Martyn, and Helen Wickstead. "'One Immense Black Spot': Aerial Views of London, 1784–1918." *London Journal* 35, no. 3 (2010): 236–54.

Barnes, Julian. *Levels of Life.* London: Jonathan Cape, 2013.

Barnes, Trevor J., and Matthew Farish. "Between Regions: Science, Militarism, and American Geography from World War to Cold War." *Annals of the Association of American Geographers* 96, no. 4 (2006): 807–826.

Barry, Roger, and Richard Chorley. *Atmosphere, Weather, and Climate.* 9th ed. London: Routledge, 2009.

Barthelme, Donald. *Sixty Stories.* New York: Penguin, 2003.

Bassett, Bruce, Bob Nichol, and Daniel Eisenstein. "Sounding the Dark Cosmos." *Astronomy and Geophysics* 46, no. 5 (2005): 526–29.

Beck, Richard. "Remote Sensing and GIS as Counterterrorism Tools in the Afghanistan War: A Case Study of the Zhawar Kili Region." *Professional Geographer* 55, no. 2 (2003): 170–79.

Bennett, Jane. *The Enchantment of Modern Life: Attachments, Crossings, Ethics.* Princeton, NJ: Princeton University Press, 2001.

———. "Systems and Things: A Response to Graham Harman and Timothy Morton." *New Literary History* 43, no. 2 (2012): 225–33.

———. *Vibrant Matter: A Political Ecology of Things.* Durham, NC: Duke University Press, 2010.

Bennett, Jane, and William Connolly. "The Crumpled Handkerchief." In *Time and History in Deleuze and Serres*, edited by Bernd Herzogenrath, 153–72. London: Continuum, 2012.

Bennett, Jill. *Practical Aesthetics: Events, Affects, and Art after 9/11*. London: I. B. Tauris, 2012.

Bennington, Geoffrey. *Interrupting Derrida*. London: Routledge, 2000.

Bergson, Henri. *The Creative Mind*. Mineola, NY: Dover, 2007.

Berland, Jody. "Remote Sensors: Canada and Space." *Semiotext(e)* 6, no. 2 (1994): 28–25.

Berlant, Lauren. "The Commons: Infrastructures for Troubling Times." *Environment and Planning D: Society and Space* 34, no. 3 (2016): 393–419.

———. *Cruel Optimism*. Durham, NC: Duke University Press, 2011.

———. "Unfeeling Kerry." *Theory and Event* 8, no. 2 (2005): n.p.

Berman, Marshall. *All That Is Solid Melts into Air: The Experience of Modernity*. London: Verso, 2010.

Biehl-Missal, Brigitte, and Michael Saren. "Atmospheres of Seduction: A Critique of Aesthetic Marketing Practices." *Journal of Macromarketing* 32, no. 2 (2012): 168–80.

"The Big Balloon in Paris." *New York Times*, August 5, 1878.

"A Big Echo's Little Pal." *Life Magazine*, June 21, 1963, 26.

Bille, Mikkel. "Lighting Up Cosy Atmospheres in Denmark." *Emotion, Space, and Society* 15, no. 1 (2015): 56–63.

Bille, Mikkel, Peter Bjerregard, and Tim Flohr Sørensen. "Staging Atmospheres: Materiality, Culture, and the Texture of the In-Between." *Emotion, Space, and Society* 15, no. 1 (2015): 31–38.

Bishop, Elizabeth. *Poems*. New York: Farrar, Straus and Giroux, 2011.

Bissell, David. "Animating Suspension: Waiting for Mobilities." *Mobilities* 2, no. 2 (2007): 227–98.

———. "Passenger Mobilities: Affective Atmospheres and the Sociality of Public Transport." *Environment and Planning D: Society and Space* 28, no. 2 (2010): 270–89.

Bogost, Ian. *Alien Phenomenology, or, What It's Like to Be a Thing*. Minneapolis: University of Minnesota Press, 2012.

Böhme, Gernot. "Atmosphere as the Fundamental Concept of a New Aesthetics." *Thesis Eleven* 36 (1993): 113–26.

Bradbury, Ray. *The Martian Chronicles*. London: Harper Collins, 2008.

———. "Take Me Home." *New Yorker*, June 4, 2012. http://www.newyorker.com /reporting/2012/06/04/120604fa_fact_bradbury. Last accessed, June 19, 2014.

Brennan, Teresa. *The Transmission of Affect*. Ithaca, NY: Cornell University Press, 2004.

Brienza, Casey. "Remembering the Future: Cartooning Alternative Life Courses in *Up* and *Future Lovers*." *Journal of Popular Culture* 46 (2013): 299–314.

Brown, Nathan. "The Technics of Prehension: On the Photography of Nicholas Baier." In *The Lure of Whitehead*, edited by Nicholas Gaskill and A. J. Nocek, 127–54. Minneapolis: Minnesota Press, 2014.

Brumfiel, Geoff. "Big Bang's Ripples: Two Scientists Recall Their Big Discovery." *NPR*, May 20, 2014. http://www.npr.org/2014/05/20/314239930/big-bangs -afterglow-two-scientists-recall-their-big-discovery. Last accessed August 14, 2015.

Bruno, Guliana. *Surface: Matters of Aesthetics, Materiality, and Media*. Chicago, IL: University of Chicago Press, 2014.

Bruun Jenson, Casper, and Atsuro Morita. "Infrastructures as Ontological Experiments." *Ethnos* 82, no. 4 (2017): 615–26.

Bryant, Levi. *Onto-Cartography: An Ontology of Machines and Media*. Edinburgh, Scotland: Edinburgh University Press, 2014.

Butler, Judith. *Precarious Life: The Powers of Mourning and Violence*. London: Verso, 2004.

Butler, Will. "Can We Trust Google with the Stratosphere?" *Atlantic* (2013). www .theatlantic.com/technology/archive/2013/08/can-we-trust-google-with-the -stratosphere/278797/. Last accessed March 7, 2016.

Campbell, James. *Introduction to Remote Sensing*. 4th ed. London: Taylor Francis, 2006.

Childs, Philip. *Texts: Contemporary Cultural Texts and Critical Approaches*. Edinburgh, Scotland: Edinburgh University Press, 2006.

Chollet, L. "Balloon Fabrics Made of Goldbeater's Skins." *L'aéronautique*, August 1922.

Choy, Timothy. *Ecologies of Comparison: An Ethnography of Endangerment in Hong Kong*. Durham, NC: Duke University Press, 2011.

Choy, Timothy, and Jerry Zee. "Condition: Suspension." *Cultural Anthropology* 30, no. 2 (2015): 210–23.

C. J. S. "A Series of Letters on Acoustics, Addressed to Mr. Alexander, Durham Palace, West Hackney." *Gentleman's Magazine* 111, no. 2 (February 1812): 105–10.

Clark, Nigel. *Inhuman Natures: Sociable Life on a Dynamic Planet*. London: Sage, 2011.

Clarke, Arthur C. "Extra-Terrestrial Relays: Can Rocket Stations Give World-Wide Radio Coverage?" *Wireless World* 51 (1945): 305–8.

Clarke, David, and Marcus Doel. "Engineering Space and Time: Moving Pictures and Motionless Trips." *Journal of Historical Geography* 31 no. 1 (2005): 41–60.

Clough, Patricia. "Praying and Playing to the Beat of a Child's Metronome." *Subjectivity* 3, no. 4 (2010): 349–65.

———. "The Transformational Object of Cruel Optimism." 2011. http://bcrw .barnard.edu/wp-content/uploads/2012/Public-Feelings-Responses/Patricia -Ticineto-Clough-The-Transformational-Object-of-Cruel-Optimism.pdf. Last accessed November 3, 2015.

Cohen, Jeffrey, and Lowell Duckert, eds. *Elemental Ecocriticism: Thinking with Earth, Air, Water, Fire*. Minneapolis: University of Minnesota Press, 2015.

Coles, Peter. *The Routledge Critical Dictionary of the New Cosmology*. New York: Routledge, 1999.

Colls, Rachel, and Maria Fannin. "Placental Surfaces and the Geographies of Bodily Interiors." *Environment and Planning A* 45, no. 5 (2013): 1087–104.

Conley, Cornelius. "The Great Japanese Balloon Offensive." *Air University Review*, January–February 1968. http://www.airpower.maxwell.af.mil/airchronicles /aureview/1968/jan-feb/conley.html. Last accessed May 26, 2014.

Connolly, William. *A World of Becoming*. Durham, NC: Duke University Press, 2011.

Connor, Steven. "Man Is a Rope." Unpublished manuscript. 2008. http://www .stevenconnor.com/rope/. Last accessed May 13, 2014.

———. *The Matter of Air: Science and Art of the Ethereal*. London: Reaktion Books, 2010.

Cook, Ian. "Follow the Thing: Papaya." *Antipode* 36, no. 4 (2004): 642–64.

Corbin, Christina. "Bible Drop: Christian Group Takes to the Sky to Sneak Bibles into North Korea." *Fox News*, November 8, 2013. http://www.foxnews.com/world /2013/11/08/bible-drop-christian-group-takes-to-the-sky-to-sneak-gospel-into -north-korea/?intcmp=latestnews. Last accessed January 9, 2014.

Cosgrove, Denis. *A Cartographic Genealogy of the Earth in the Western Imagination*. Baltimore, MD: Johns Hopkins University Press, 2001.

Couture, Jean-Pierre. "Spacing Emancipation? Or How Sphereology Can Be Seen as a Therapy for Modernity." *Environment and Planning D: Society and Space* 27, no. 1 (2009): 157–63.

Crampton, Jeremy. "Cartographic Calculations of Territory." *Progress in Human Geography* 35, no. 1 (2010): 92–103.

Crandall, Jordan. "Summary of *Gatherings*." (2011) http://jordancrandall.com/main /+GATHERINGS/index.html. Last accessed November 12, 2015.

Cresswell, Tim, and Craig Martin. "On Turbulence: Entanglements of Disorder and Order on a Devon Beach." *Tijdschrift Voor Economische en Sociale Geographie* 103 (2012): 516–29.

Cummings, E. E. *Selected Poems*. New York: Norton, 1994.

Cummings, Richard H. *Cold War Radio: The Dangerous History of American Broadcasting in Europe, 1950–1989*. Jefferson, NC: McFarland, 2009.

———. *Radio Free Europe's "Crusade for Freedom": Rallying Americans behind Cold War Broadcasting, 1950–1960*. Jefferson, NC: McFarland, 2010.

Cupples, Julie. "Culture, Nature and Particulate Matter—Hybrid Reframings in Air Pollution Scholarship." *Atmospheric Environment* 43, no. 1 (2009): 207–17.

Damjanov, Katarina. "The Matter of Media in Outer Space: Technologies of Cosmobiopolitics." *Environment and Planning D: Society and Space* 33, no. 5 (2015): 889–906.

Davies, Merton E., and William R. Harris. "RAND's Role in the Evolution of Balloon and Satellite Observation Systems, and Related US Space Technology." Santa Monica, CA: RAND Corporation, R-3692-RC, 1988.

de Bernardis P., et al. "Images of the Early Universe from the BOOMERanG experiment." In J. C. Wheeler and H. Martel eds, *Relativistic Astrophysics*. AIP Conference Proceedings. No.586. Melville, N.Y: American Institute of Physics, 2011, 157–71

Deleuze, Gilles. *Difference and Repetition.* Translated by Paul Patton. London: Continuum, 1994.

———. *Expressionism in Philosophy: Spinoza.* Translated by Martin Joughin. New York: Zone Books, 1990.

———. *The Fold.* Translated by Tom Conley. London: Athlone, 1993.

———. *Immanence: Essays on a life.* Translated by Anne Boyman. New York: Zone Books, 2001.

Deleuze, Gilles, and Félix Guattari. *A Thousand Plateaus.* Translated by Brian Massumi (Minneapolis: University of Minnesota Press, 1987).

DeSilvey, Caitlin. "Observed Decay: Telling Stories with Mutable Things." *Journal of Material Culture* 11, no. 3 (2006): 318–38.

Derrida, Jacques. *Specters of Marx: The State of the Debt, the Work of Mourning, and the New International.* Translated by Peggy Kamuf. New York: Routledge, 1994.

Dessauce, Marc. *The Inflatable Moment: Pneumatics and Protest in '68.* New York: Princeton Architectural Press, 1999.

de Terzi, Francesco Lana. *The Aerial Ship.* London: Aeronautical Society of Great Britain, 1910.

Donald, Mark. "Balloon-Borne Messages to North Korea Have Detractors on Both Sides of the Border." *New York Times*, April 26, 2011. http://www.nytimes.com /2011/04/27/world/asia/27iht-korea.html?pagewanted=all&_r=0Mark. Last accessed January 9th 2014.

Dorrian, Mark, and Frédéric Pousin, eds. *Seeing from Above: The Aerial View in Visual Culture.* London: IB Tauris, 2013.

Drapeau, Raoul, E. "Operation Outward: Britain's World War II Offensive Balloons." *IEEE Power and Energy Magazine*, September/October (2011): 94–105.

Duff, Cameron. "Atmospheres of Recovery: Assemblages of Health," *Environment and Planning A* 48, no. 1 (2016), 58–74.

Durham Peters, John. *The Marvelous Clouds: A Philosophy of Elemental Media.* Chicago: University of Chicago Press, 2015.

Dyson, Frances. *The Tone of Our Times: Sound, Sense, Economy, and Ecology.* Cambridge, MA: MIT Press, 2014.

———. *Sounding New Media: Immersion and Embodiment in the Arts and Culture.* Berkeley: University of California Press, 2009.

Edensor, Tim and Shanti Sumartajo. "Designing Atmospheres: Introduction to Special Issue." *Visual Communication* 14, no. 3 (2015): 251–65.

Ehrlich, Heyward. "Poe in Cyberspace: Balloons! Drones!! The Global Internet!!!." *The Edgar Allan Poe Review* 16, no. 2 (2015): 242–46.

Ekström, Anders. "Seeing From Above: A Particular History of the General Observer." *Nineteenth Century Contexts* 31, no. 3 (2009): 185–207.

Elden, Stuart. "Secure the Volume: Vertical Geopolitics and the Depth of Power." *Political Geography* 34, no. 1 (2013): 35–51.

Ellis, Patrick. "'Panoramic Whew': The Aeroscope at the Panama-Pacific International Exposition, 1915." *Early Modern Popular Visual Culture* 13, no. 3 (2015): 209–31.

Engelmann, Sasha. *The Cosmological Aesthetics of Tomás Saraceno's Atmospheric Experiments*. DPhil Thesis, Oxford University, 2017.

———. "More-Than-Human Affinitive Listening." *Dialogues in Human Geography* 5, no. 1 (2015): 76–79.

———. "Toward a Poetics of Air: Sequencing and Surfacing Breath." *Transactions of the Institute of British Geographers* 40, no. 3 (2015): 430–44.

Engelmann, Sasha, and Derek McCormack. "Elemental Aesthetics: On Artistic Experiments with Solar Energy." *Annals of the Association of American Geographers* 108, no. 1 (2018): 241–59.

Engelmann, Sasha, Derek McCormack, and Bronislaw Szerszynski. "Becoming Aerosolar and the Politics of Elemental Association." In *Becoming Aerosolar*, by Tomás Saraceno, 67–101. Vienna: 21er Haus, 2015.

Engels-Schwarzpaul, A-Chr. "A Warm Grey Fabric Lined on the Inside with the Most Lustrous and Colourful of Silks: Dreams of Airships and Tropical Islands." *Journal of Architecture* 12, no. 5 (2007): 525–42.

Erben, Walter. *Marc Chagall*. New York: F. A. Praeger, 1966.

Erdman, David, Marcelyn Gow, Ulrika Karlsson, and Chris Perry. "Parallel Processing: Design/Practice." *Architectural Design* 76 (2006): 80–87.

Experiments in Art and Technology (EAT). "The Great Big Mirror Dome Project." *Experiments in Art and Technology Los Angeles Records, 1969–1979*. Los Angeles: The John Paul Getty Trust, 1970.

Faber, Roland, and Andrew Goffey, eds. *The Allure of Things*. London: Bloomsbury, 2014.

Flammarion, Camille. *L'atmosphère: Météorologie populaire*. Paris: Hachette, 1888.

Flusser, Vilém. *The Shape of Things: A Philosophy of Design*. Translated by Anthony Mathews. London: Reaktion Books, 1999.

Flynn, Kate. "Fat and the Land: Size Stereotyping in Pixar's *Up*." *Children's Literature Association Quarterly* 35, no. 4 (2010): 435–42.

Forsyth, Isla, Hayden Lorimer, Peter Merriman, and James Robinson. "What Are Surfaces?" *Environment and Planning A* 45, no. 5 (2013): 1013–20.

Forsythe, William. *Suspense: Exhibition Catalogue*. Zurich: JRP/ Ringier Kunstverlag, 2008.

Foster, Hal. "Polemics, Postmodernism, Immersion, Militarized Space." *Journal of Visual Culture* 3, no. 3 (2004): 320–35.

Freedgood, Elaine. *Victorian Writing about Risk: Imagining a Safe England in a Dangerous World*. Cambridge: Cambridge University Press, 2000.

Freyssinet, Eugène. "On the Sublime." *Architectural Research Quarterly* 5 (2001): 249–53.

Gabrys, Jennifer. *Environmental Sensing Technology and the Making of a Computational Planet*. Minneapolis: University of Minnesota Press, 2016.

Gallagher, Michael. "Field Recording and the Sounding of Spaces." *Environment and Planning D: Society and Space* 33, no. 3 (2015): 560–76.

Gallagher, Michael, Anja Kanngieser, and Jonathan Prior. "Listening Geographies: Landscape, Affect and Geotechnologies." *Progress in Human Geography* 41, no. 5 (2017): 618–37.

Galloway, Andrew, and Eugene Thacker. *The Exploit: A Theory of Networks.* Minneapolis: University of Minnesota Press, 2007.

Gaskill, Nicholas, and A. J. Nocek, eds. *The Lure of Whitehead.* Minneapolis: University of Minnesota Press, 2014.

Gavaghan, Helen. *Something New Under the Sun: Satellites and the Beginning of the Space Race.* New York: Springer, 1998.

Gertner, Jon. *The Idea Factory: Bell Labs and the Great Age of American Invention.* London: Penguin, 2012.

"GHOST Balloon Completes One-Year Flight." *National Center for Atmospheric Research* (Boulder, Colorado), October 1, 1968, 1. http://nldr.library.ucar.edu /repository/assets/info/INFO-000–000–000–105.pdf. Last accessed June 4, 2014.

Giblet, M. A. "The International Free-Balloon Race for the Gordon Bennett Cup." *Quarterly Journal of the Royal Meteorological Society* 50 (1924): 260–67.

Giessen, David. *Manhattan Atmospheres: Architecture, the Interior Environment, and Urban Crisis.* Minneapolis: University of Minnesota Press, 2013.

Gillespie, Charles Coulston. *Science and Polity in France: The End of the Old Regime.* Princeton, NJ: Princeton University Press, 2004.

Gins, Madeleine, and Arakawa. *Architectural Body.* Tuscaloosa: Alabama University Press, 2002.

Glaisher, James, ed. *Travels in the Air.* 2nd ed. London: Richard Bentley and Son, 1871.

Goody, R. M. *The Physics of the Stratosphere.* Cambridge: Cambridge University Press, 1958.

"Google Installing Hundreds of Internet-Enabled Balloons in Indonesia." *Guardian*, October 28, 2015. http://www.theguardian.com/technology/2015/oct/28 /google-installing-20000-internet-enabled-balloons-in-indonesia. Last accessed March 8, 2016.

Gratton, Peter. *Speculative Realism: Prospects and Problems.* London: Bloomsbury, 2014.

Gregory, Derek. "From a View to a Kill: Drones and Late Modern War." *Theory, Culture, and Society* 28, nos. 7–8 (2011): 188–215.

———. "Moving Targets and Violent Geographies." In *Spaces of Danger: Culture and Power in the Everyday*, edited by Heather Merrill and Lisa Hoffman, 256–97. Athens: University of Georgia Press, 2015.

Griffero, Tonino. *Atmospheres: Aesthetics of Emotional Spaces.* Farnham, UK: Ashgate, 2014.

Groening, Stephen. "Towards a Meteorology of the Media." *Transformations* 25 (2014): n.p.

Guattari, Félix. *Chaosmosis: An Ethico-Aesthetic Paradigm.* Translated by Paul Bains and Julian Pefanis. Sydney: Power, 1995.

Gumbrecht, Hans Ulrich. *Atmosphere, Mood, Stimmung: On a Hidden Potential of Literature*. Translated by Erik Butler. Stanford, CA: Stanford University Press, 2012.

Guth, Alan. *The Inflationary Universe: The Quest for a New Theory of Cosmic Origins*. London: Jonathan Cape, 1997.

Hall, Rex, and David Shayler. *The Rocket Men: Vostok and Voskhod, the First Soviet Manned Spaceflight*. Chichester, UK: Springer Praxis, 2001.

Haining, Peter, ed. *The Dream Machines: An Eye-Witness History of Ballooning*. London: New English Library, 1972.

Hallam, Elizabeth, and Jenny Hockey. *Death, Memory, and Material Culture*. Oxford: Berg, 2001.

Hallion, Richard. *Taking Flight: Inventing the Aerial Age, from Antiquity through the First World War*. Oxford: Oxford University Press, 2003.

Hallward, Peter. *Out of This World: Deleuze and the Philosophy of Creation*. London: Verso, 2006.

Hansen, Mark. *Feed-Forward: On the Future of Twenty-First-Century Media*. Chicago, IL: University of Chicago Press, 2015.

Haraway, Donna. "The Persistence of Vision." In *The Visual Culture Reader*, edited by Nicholas Mirzoeff, 191–98. London: Routledge 2001.

———. *Simians, Cyborgs, and Women: The Reinvention of Nature*. London: Routledge, 1991.

———. "Situated Knowledges: The Science Question in Feminism and the Privilege of Partial Perspective." *Feminist Studies*, 14, no. 3 (1988): 575–599.

Hare, F. K. "The Stratosphere." *Geographical Review* 52, no. 1 (1962): 525–47.

Harman, Graham. *Guerrilla Metaphysics: Phenomenology and the Carpentry of Things*. Peru, IL: Open Court, 2010.

———. "Realism without Materialism." *SubStance* 40 (2011): 52–72.

———. *Towards Speculative Realism: Essays and Lectures*. Ropley, UK: Zero Books, 2010.

———. "Whitehead and Schools X, Y, and Z." In *The Lure of Whitehead*, edited by Nicholas Gaskill and A. J. Nocek, 231–48. Minneapolis: University of Minnesota Press, 2014.

Harp, Stephen. *Marketing Michelin: Advertising and Cultural Identity in Twentieth-Century France*. Baltimore, MD: Johns Hopkins University Press, 2001.

Harrison, Paul. "*Flētum*: A Prayer for X." *Area* 43 (2011): 158–61.

Harvey, Penny, et al., eds. *Objects and Materials: A Routledge Companion*. London: Routledge, 2014.

Hauge, Bettina. "The Air from Outside: Getting to Know the World through Air Practices." *Journal of Material Culture* 18, no. 2 (2013): 171–87.

Hawkins, Harriet, and Elizabeth Straughan. "Nano-Art, Dynamic Matter, and the Sight/Sound of Touch." *Geoforum* 51 (2014): 130–39.

Healy, Stephen. "Atmospheres of Consumption: Shopping as Involuntary Vulnerability." *Emotion, Space, and Society*, 10, no. 1 (2014): 35–43.

Heath, T. L. *The Works of Archimedes*. Cambridge: Cambridge University Press, 1897.

Hecht, Gabrielle. *Nuclear Being: Africans and the Global Uranium Trade*. Cambridge, MA: MIT Press, 2012.

Hervieux, Linda. *Forgotten: The Untold Story of D-Day's Black Heroes*. New York: Harper, 2016.

Hill, Sheryl. "'Until I Have Won': Vestiges of Coverture and the Invisibility of Women in the Twentieth Century: A Biography of Jeannette Ridlon Piccard." PhD thesis, Ohio University, 2009.

Hirsh, Elizabeth, and Gary Olsen. "'Je-Luce Irigaray': A Meeting with Luce Irigaray." Translated by Elizabeth Hirsh and Gaëtan Brulotte. *Journal of Advanced Composition* 16, no. 3 (1996): 341–61.

Hoffman, Frank, Frederick Augustyn, and Martin Manning. *Dictionary of Toys and Games in American Popular Culture*. New York: Routledge, 2004.

Hoffman, Phillip. *Wings of Madness: Albert Santos-Dumont and the Invention of Flight*. New York: Theia, 2003.

Holbrow, Charles, and Joseph Amato. "What Gay-Lussac Didn't Tell Us." *American Journal of Physics* 79, no. 1 (2011): 17–24.

Holland, Keating, and Adam Levy. "Deflating: Democrats Not into Balloon Drop Tradition." CNN *Political Ticker*, September 6, 2012. http://politicalticker.blogs.cnn.com/2012/09/06/deflating-democrats-not-into-balloon-drop-tradition/. Last accessed October 24, 2016.

Holmes, Richard. *Falling Upwards: How We Took to the Air*. London: William Collins, 2013.

Howe, Cymene, and Dominic Boyer. "Aeolian Politics." *Distinktion: Scandinavian Journal of Social Theory* 16, no. 1 (2015): 31–48.

Huhtamo, Erkki. "Aeronautikon! Or, The Journey of the Balloon Panorama." *Early Popular Visual Culture* 7, no. 3 (2009): 295–306.

Huss, George Martin. "The Gas-Tank Nuisance." *Art World* 1, no. 1 (1916): 66–67.

"In Praise of Gasometers." Editorial. *Guardian*, July 20, 2011. http://www.theguardian.com/commentisfree/2011/jul/20/in-praise-of-gasometers. Last accessed April 29, 2014.

Ingold, Tim. *Being Alive*. London: Routledge, 2015.

———. *The Life of Lines*. London: Routledge, 2015.

———. *Making: Anthropology, Art, and Architecture*. London: Routledge, 2013.

———. "Footprints through the Weather-World: Walking, Breathing, Knowing." *Journal of the Royal Anthropological Institute* 16, no.1 (2010): 121–39.

———. "Materials against Materiality." *Archaeological Dialogues* 14, no. 1 (2007): 1–16.

———. *The Perception of the Environment*. London: Routledge, 2000.

Ingram, Penelope. "From Goddess Spirituality to Irigaray's Angel: The Politics of the Divine." *Feminist Review* 66 (autumn 2000): 46–72.

Irigaray, Luce. *An Ethics of Sexual Difference*. Translated by Carolyn Burke and Gillian C. Gill. London: Continuum, 2005.

———. *The Forgetting of Air in Martin Heidegger*. Translated by Mary Beth Mader. London: Athlone, 1999.

Irwin, Robert. *Being and Circumstance: Notes towards a Conditional Art*. San Francisco, CA: Lapis Press, 1985.

Jackson, Mark, and Maria Fannin. "Letting Geography Fall Where It May— Aerographies Address the Elemental." *Environment and Planning D: Society and Space* 29, no. 3 (2011): 435–44.

Jakes, William. *Participation of Bell Telephone Laboratories in Project Echo and Experimental Results: Technical Note D-1127*. Washington, DC: NASA, 1961.

Jellis, Thomas, and Joe Gerlach. "Micropolitics and the Minor." *Environment and Planning D: Society and Space* 35, no. 4 (2017): 563–67.

Jenson, Casper Bruun, and Atsuro Morita. "Infrastructures as Ontological Experiments." *Ethnos* 82, no. 4 (2017): 615–26.

Jirat-Wasiutynski, Vojtech. "The Balloon as Metaphor in the Early Work of Odilon Redon." *Artibus et Historiae* 13 (1992): 195–206.

John, George. "Welcome to the Space Jam: How United States Regulators Should Govern Google and Facebook's New Internet-Providing High-Altitude Platforms." *American University Business Law Review* 471 (2015): 471–504.

Johnson, Vida T., and Graham Petrie. *The Films of Andrei Tarkovksy: A Visual Fugue*. Bloomington: Indiana University Press, 2004.

Jones, Jonathan. "Let There Be Light." *Observer*, March 19, 2000.

Jones, Rachel. "Vital Matters and Generative Materiality: Between Bennett and Irigaray." *Journal of the British Society for Phenomenology* 46, no. 22 (2015): 156–72.

Joselit, David. "Yippie Pop: Abbie Hoffman, Andy Warhol, and Sixties Media Politics." *Grey Room* 8 (2002): 62–79.

Kanngieser, Anja. "A Sonic Geography of Voice: Towards an Affective Politics." *Progress in Human Geography* 36, no. 3 (2012): 336–53.

Kaplan, Caren. *Aerial Aftermaths: Wartime From Above*. Durham and London: Duke University Press, 2018.

———"The Balloon Prospect." In *From Above: War, Violence, and Verticality*, edited by Peter Adey, Mark Whitehead, and Alison Williams, 19–40 London: Hurst, 2013.

———. "Mobility and War: The Cosmic View of US Air Power." *Environment and Planning A* 38, no. 2 (2006): 395–407.

———. "'A Rare and Chilling View': Aerial Photography as Biopower in the Visual Culture of '9/11.'" *Reconstruction* 11, no. 2 (2011), np.

Keen, Paul. "The 'Balloonomania': Science and Spectacle in 1780s England." *Eighteenth-Century Studies* 39, no. 4 (2006): 507–53.

Kim, Mi Gyung. "Balloon Mania: News in the Air." *Endeavour* 28, no. 4 (2004): 149–55.

Klüver, Billy, ed. *Pavilion: Experiments in Art and Technology*. New York: Dutton, 1972.

———. "Photographic Recording of Some Optical Effects in a 27.5m Spherical Mirror." *Applied Optics* 10, no. 12 (1971): 27–55.

Knight, Laura. *The Magic of a Line*. London: Kimber, 1965.

Koddenberg, Matthias. "'It Boggles the Mind': The Air Packages of Christo and Jeanne-Claude." In *Christo: Big Air Package*, edited by Wolfgang Volz and Peter Pachinke, 131–43. Essen: Klartext, 2013.

Lachambre, Henri, and Alexis Machuron. *Andrée's Balloon Expedition in Search of the North Pole*. New York: Frederick Stokes, 1898.

Lappin, Yakkov. "A Common Sight during the Gaza War, IDF's Reliance on Aerostat Balloons Is Up." *Jerusalem Post*, June 10, 2014. http://www.jpost.com /Israel-News/New-Tech/IDFs-reliance-on-aerostats-is-up-378127. Last accessed October 16, 2016.

Latour, Bruno. "Network Theory / Networks, Societies, Spheres: Reflections of an Actor-Network Theorist." *International Journal of Communication* 5 (2011): 796–810.

———. "Some Experiments in Art and Politics." *E-Flux Journal*, March 2011. http:// www.e-flux.com/journal/some-experiments-in-art-and-politics/. Last accessed September 11, 2017.

Lefebvre, Henri. *The Production of Space*. Translated by D. Nicholson-Smith. Oxford: Blackwell, 1992.

Levin, Laura. *Performing Ground: Space, Camouflage, and the Art of Blending In*. Manchester, UK: Palgrave Macmillan, 2014.

Long, David. "CBP's Eyes in the Sky: CBP's Tethered Aerostats Keep Watch for Trouble from 10,000ft." *U.S. Customs and Border Protection*, September 13, 2016. https://www.cbp.gov/frontline/frontline-november-aerostats. Last accessed October 16, 2016.

Lorimer, Jamie. "Nonhuman charisma." *Environment and Planning D: Society and Space* 25, no. 5 (2007): 911–932.

Lorimer, Jamie, Timothy Hodgetts, and Maan Barua, "Animals' Atmospheres." *Progress in Human Geography* (forthcoming).

Luciano, Dana. *Arranging Grief: Sacred Time and the Body in Nineteenth-Century America*. New York: New York Press, 2007.

Lunardi, Vincenzo. *An Account of the First Aerial Voyage in England: In a Series of Letters to His Guardian, Chevalier Gherardo Compagni, Written under the Impressions of the Various Events That Affected the Undertaking*. London: Bell, 1784.

Lynn, Michael. *The Sublime Invention: Ballooning in Europe, 1783–1820*. London: Pickering and Chatto, 2010.

Macauley, David. *Elemental Philosophy: Earth, Air, Fire, and Water as Environmental Ideas*. Albany, NY: SUNY Press, 2010.

MacDonald, Fraser. "Perpendicular Sublime: Regarding Rocketry and the Cold War." In *Observant States: Geopolitics and Visual Cultures*, edited by Fraser Mac-Donald, Rachel Hughes, and Klaus Dodds, 267–90. London: I. B. Tauris, 2010.

Mackrell, Judith. "The Joy of Sets." *Guardian*, June 6, 2005. http://www .theguardian.com/stage/2005/jun/06/dance. Last accessed November 16, 2015.

Manning, Erin. *Politics of Touch: Movement, Sense, Sovereignty*. Minneapolis: University of Minnesota Press, 2007.

————. "Propositions for the Verge: William Forsythe's Choreographic Objects." *Inflexions*, no. 2 (2009): n.p.

————. *Relationscapes: Movement, Art, Philosophy*. Cambridge, MA: MIT Press, 2009.

Marres, Noortje. *Material Participation: Technology, the Environment, and Everyday Publics*. Basingstoke, UK: Palgrave Macmillan, 2012.

Martin, Craig. "Fog-Bound: Aerial Space and the Elemental Entanglements of Body-with-World." *Environment and Planning D: Society and Space* 29, no. 3 (2011): 454–68.

Masco, Joseph. *The Theatre of Operations: National Security Affect from the Cold War to the War on Terror*. Durham, NC: Duke University Press, 2014.

Mason, Monck. *Aeronautica; Or, Sketches Illustrative of the Theory and Practice of Aerostation: Comprising an Enlarged Account of the Late Aerial Expedition to Germany*. London: F. C. Westley, 1838.

Massumi, Brian. *Ontopower: War, Powers, and the State of Perception*. Durham, NC: Duke University Press, 2015.

————. *Parables for the Virtual: Movement, Affect, Sensation*. Durham, NC: Duke University Press, 2002.

McConville, David. "Cosmological Cinema: Pedagogy, Propaganda, and Perturbation in Early Dome Theatres." *Technoetic Arts* 5, no. 2 (2007): 69–85.

McCormack, Derek. "Stratospheric Envelopes for an Atmospheric Mode of Address," *GeoHumanities*, forthcoming, http://dx.doi.org/10.1080/2373566X.2017.1301214

————. "Devices for Doing Atmospheric Things." In *Nonrepresentational Methodologies: Re-Envisioning Research*, edited by Phillip Vannini, 89–111. London: Routledge, 2015.

————. "Engineering Affective Atmospheres on the Moving Geographies of the 1897 Andrée Expedition." *Cultural Geographies* 15, no. 4 (2008): 413–30.

————. "Piloting the Aerocene." In *Aerocene Newspaper*, publication accompanying Tomás Saraceno's *Aerocene* installation at the Grand Palais, Paris, December 2015.

————. *Refrains for Moving Bodies: Experience and Experiment in Affective Spaces*. Durham, NC: Duke University Press, 2013.

————. "Remotely Sensing Affective Afterlives: The Spectral Geographies of Material Remains." *Annals of the Association of American Geographers* 100, no. 3 (2010): 640–54.

McCullough, David. *The Great Bridge: The Epic Story of the Building of the Brooklyn Bridge*. New York: Simon and Schuster, 1972.

McCullough, Malcolm. *Ambient Commons: Attention in the Age of Embodied Information*. Cambridge, MA: MIT Press, 2013.

McEwan, Ian. *Enduring Love*. London: Jonathan Cape, 1997.

Meikle, Jeffrey. *American Plastic: A Cultural History*. New Brunswick, NJ: Rutgers University Press, 1997.

Metcalfe, Andrew. "Nothing: The Hole, the Holy, the Whole." *Ecumene* 8, no. 3 (2001): 247–63.

Miller, Paul. "The Engineer as Catalyst: Billy Klüver on Working with Artists." *Spectrum, IEEE* 35, no. 7 (1998): 20–29.

Morden, Barbara. *Laura Knight: A Life*. Pembroke, UK: McNidder and Grace, 2014.

Morrison, Carey-Anne, Lynda Johnston, and Robyn Longhurst. "Critical Geographies of Love as Spatial, Relational and Political." *Progress in Human Geography* 37, no. 4 (2013): 505–21.

Morton, Timothy. *Ecology without Nature: Rethinking Environmental Aesthetics*. Cambridge, MA: Harvard University Press, 2007.

———. "Here Comes Everything: The Promise of Object-Oriented Ontology." *Qui Parle: Critical Humanities and Social Sciences* 19, no. 2 (2011): 163–90.

———. *Hyperobjects: Philosophy and Ecology after the End of the World*. Minneapolis: University of Minnesota Press, 2013.

Nadar (Gaspard-Félix Tournachon). "My Life as Photographer." Translated by Thomas Repensek. *October* 5 (1978): 6–28.

Nancy, Jean-Luc. *The Inoperative Community*. Edited by Peter Connor. Translated by Peter Connor, Lisa Garbus, Michael Holland, and Sinoma Sawhney. Minneapolis: University of Minnesota Press, 1991.

Nieuwenhuis, Marjin. "Atmospheric Governance: Gassing as law for the protection and killing of life." *Environment and Planning D: Society and Space*, forthcoming, https://doi.org/10.1177/0263775817729378

"The Odyssey of Project Echo." *SP-4308 Spaceflight Revolution*. Washington, DC: NASA History Office, n.d.

Ohrstedt, Pernilla. "Feeling Reflective." *RIBJA Newsletter*, March 1, 2013. https://www.ribaj.com/culture/feeling-reflective. Last accessed December 20, 2016.

"100,000 Children See Store Pageant: Macy's Christmas Parade Is Greeted by Cheers along Broadway." *New York Times*, November 30, 1928, 25.

Oreskes, Naomi and John Krige eds. *Science and Technology in the Global Cold War* Cambridge, MA: MIT Press, 2014.

Needell, Allan. "Truth Is Our Weapon": Project TROY, Political Warfare, and Government-Academic Relations in the National Security State." *Diplomatic History* 17, no. 3 (1993): 399–420.

Packer, Randall. "The Pavilion: Into the 21st Century: A Space for Reflection." *Organized Sound* 9, no. 3 (2004): 251–59.

Palmer, Michaela, and Owain Jones. "On Breathing and Geography: Explorations of Data Sonifications of Timespace Processes with Illustrating Examples from a Tidally Dynamic Landscape (Severn Estuary, UK)." *Environment and Planning A* 46, no. 1 (2014): 222–40.

"Park Sang-Hak on Receiving the Václav Havel Prize for Creative Dissent." *News Focus International*, May 7, 2013. http://newfocusintl.com/park-sang-hak-vaclav-havel-prize-for-creative-dissent/. Last accessed January 9, 2014.

Parikka, Jussi. *A Geology of Media*. Minneapolis: University of Minnesota Press, 2015.

Parks, Lisa. *Cultures in Orbit: Satellites and the Televisual*. Durham, NC: Duke University Press, 2005.

Parks, Lisa, and Caren Kaplan, eds. *Life in the Age of Drone Warfare*. Durham, NC: Duke University Press, 2017.

Pearman, Hugh. *Airports: A Century of Architecture*. London: Lawrence King, 2004.

Pedlow, Gregory, and Donald Welzenbach. *The Central Intelligence Agency and Overhead Reconnaissance: The U-2 and OXCART Programs, 1954–1974*. Washington, DC: Central Intelligence Agency, 1992.

Penzias, Arno, and Robert Wilson. "A Measurement of Excess Antenna Temperature at 4080 Mc/s." *Astrophysical Journal* 142 (1965): 419–21.

Peterson, Andreas, Craig Timberg, and Christian Davenport. "The Military Lost Control of a Giant Unmanned Surveillance Blimp." *Washington Post*, October 28, 2015. https://www.washingtonpost.com/news/the-switch/wp/2015/10/28/the-army-lost-control-of-a-giant-unmanned-surveillance-blimp/. Last accessed November 10, 2015.

Phelan, Peggy. *Unmarked: The Politics of Performance*. London: Routledge, 1993.

Philippopoulos-Mihalopoulos, Andreas. *Spatial Justice: Body, Lawscape, Atmosphere*. London: Routledge, 2015.

———. "Withdrawing from Atmosphere: An Ontology of Air Partitioning and Affective Engineering." *Environment and Planning D: Society and Space* 34, no. 1 (2016): 150–67.

Pierce, John. "Satellite Science and Technology." *Observatory* 83 (1963): 207–16.

Pickles, John. ed. *Ground Truth: The Social Implications of GIS*. New York: Guildford, 1995.

———. *A History of Spaces: Cartographic Reason, Mapping, and the Geo-Coded World*. London: Routledge, 2003.

Pinkerton, Alastair, Stephen Young, and Klaus Dodds. "Postcards from Heaven: Critical Geographies of the Cold War Military-Industrial-Academic Complex." *Antipode* 43, no. 3 (2011): 820–44.

Pisters, Patricia. *The Neuro-Image: A Deleuzian Film-Philosophy of Digital Screen Culture*. Stanford, CA: Stanford University Press, 2012.

Pommereau, Jean-Pierre. "Observation Platforms: Balloons." In *Encyclopedia of Atmospheric Sciences*, edited by J. Holton, J. Pyle, and J. A. Curry, 1429–38. Waltham, MA: Academic Press, 2003.

Popper, Ben. "Inside Project Loon: Google's Internet in the Sky Is Almost Open for Business." *Verge*, March 2, 2015. http://www.theverge.com/2015/3/2/8129543/google-x-internet-balloon-project-loon-interview. Last accessed September 9, 2016.

"Preliminary Design of an Experimental World-Circling Spaceship." Report no. SM-11827. Santa Monica, CA: Douglas Aircraft Company, 1946.

"Project Loon: Google Balloon That Beams Down Internet Reaches Sri Lanka." *Guardian*, February 16, 2016. http://www.theguardian.com/technology/2016/feb/16/project-loon-google-balloon-that-beams-down-internet-reaches-sri-lanka. Last accessed March 9, 2016.

Rawes, Peg. *Irigaray for Architects*. London: Routledge, 2007.

Revill, George. "How Is Space Made in Sound? Spatial Mediation, Critical Phenomenology and the Political Agency of Sound." *Progress in Human Geography*, 40, no. 2 (2016): 240–56.

Richards, Lee. "The 'M'-Balloon Unit: British Balloon Distribution of Aerial Propaganda during WWII." http://www.psywar.co.uk/psywar/reproductions/mballoon.pdf. Last accessed May 26, 2014.

Riches, Alan. "Balloons: What Have They Ever Done for Us? The Contribution of the Balloon to the History of Air Power." *Royal Air Force Historical Society Journal* 27 (2002): 24–42.

Robbins, John. "Up in the Air: Balloonomania and Scientific Performance." *Eighteenth-Century Studies* 48, no. 4 (2015): 521–38.

Roberts, Tom. "From Things to Events: Whitehead and the Materiality of Process." *Environment and Planning D: Society and Space* 32, no. 6 (2014): 968–98.

Rolt, L.T.C. *The Balloonists: A History of the First Aeronauts.* Stroud, UK: Sutton, 2006.

Rose, Gillian. *Feminism and Geography: On the Limits of Geographical Knowledge.* Minneapolis: University of Minnesota Press, 1993.

Rose, Mitch. "Gathering Dreams of Presence: A Project for the Cultural Landscape." *Environment and Planning D: Society and Space* 24, no. 4 (2006): 537–54.

Salisbury, Laura. "Michel Serres: Science, Fiction, and the Shape of Relation." *Science Fiction Studies* 33, no. 1 (2006): 30–52.

Saraceno, Tomás. *Cloud Cities.* Edited by Marion Ackerman, Daniel Birnbaum, Udo Kittelmann, and Hans Ulrich Obrist. Berlin: Distanz Verlag, 2011.

Saraceno, Tomás, Sasha Engelmann, and Bronislaw Szerszynski. "Becoming Aerosolar: From Solar Sculptures to Cloud Cities." In *Art in the Anthropocene: Encounters among Aesthetics, Politics, Environments, and Epistemologies*, edited by Heather Davis and Etienne Turpin, 57–62. London: Open Humanities Press, 2015.

"Saw Wife and Five Children Killed by Jap Balloon." *Seattle Times*, June 1 1945.

Schmidt, Ulrik. "Ambience and Ubiquity." In *Throughout: Art and Culture Emerging with Ubiquitous Computing*, edited by Ulrik Ekman, 175–88. Cambridge, MA: MIT Press, 2013.

Schneemann, Peter. "Monumentalism as a Rhetoric of Impact." *Anglia* 131, nos. 2–3 (2013): 282–96.

Sculle, Keith, and John A. Jakle. "Signs in Motion: A Dynamic Agent in Landscape and Place." *Journal of Cultural Geography* 25, no. 1 (2008): 57–85.

Serres, Michel. *Angels: A Modern Myth.* Translated by Francis Cowper. Paris: Flammarion, 1995.

———. *Biogea.* Translated by Randolph Burks. Minneapolis, MN: Univocal, 2012.

———. *The Five Senses: A Philosophy of Mingled Bodies.* Translated by Margaret Sankey and Peter Cowley. London: Continuum, 2008.

———. *Genesis.* Translated by Genevieve James and James Nielson. Ann Arbor: University of Michigan Press, 1995.

———. "Jules Verne's Strange Journeys." *Yale French Studies* 52 (1975): 174–88.

———. *Rameaux*. Paris: Editions Le Pommier, 2007.

Serres, Michel, George Diez, and Christopher Roth. "The Neolithic Age Is Over." *032c* 20. 2010/2011. http://www.christopherroth.org/?p=2799. Last accessed September 11, 2017.

Shapiro, Nick. "Alter-Engineered World." In *Aerocene Newspaper*, publication accompanying Tomás Saraceno's *Aerocene* installation at the Grand Palais, Paris, December 2015.

Shaviro, Steven. *The Universe of Things: On Speculative Realism*. Minneapolis: University of Minnesota Press, 2014.

———. *Post-Cinematic Affect*. Hants, UK: O-Books, 2010.

Shaw, Ian. *Predator Empire: Drone Warfare and Full Spectrum Dominance*. Minneapolis: University of Minnesota Press, 2016.

Sheller, Mimi. *Aluminum Dreams: The Making of Light Modernity*. Cambridge, MA: MIT Press, 2014.

Sheren, Nicole Ile. *Portable Borders: Performance Art and Politics on the Frontera since 1984*. Austin: University of Texas Press, 2015.

Siegel, Marcia. "Dancing in the Dust." *Hudson Review* 54, no. 3 (2001): 455–63.

Silk, Gerald. "Myth and Meanings in Manzoni's *Merda d'Artista*." *Art Journal* 52, no. 3 (1993): 65–75.

Simone, AbdouMaliq. "Passing Things Along: (In)completing Infrastructure." *New Diversities* 17, no. 2 (2015): 151–62.

Singh, S. H. "From Hunters to Balloon Sellers," *The Hindu*, April 19, 2010. http://www.hindu.com/2010/04/19/stories/2010041954190200.htm. Last accessed November 3, 2015.

"16 Get Balloon Prizes: Fall River Boy Receives Macy's Silver Trophy." *New York Times*, December 27, 1930, 30.

Skof, Lenart, and Emily Holmes, eds. *Breathing with Luce Irigaray*. London: Bloomsbury, 2013.

Sloterdijk, Peter. "Airquakes." *Environment and Planning D: Society and Space* 27, no. 1 (2009): 41–57.

———. *Bubbles: Spheres*. Vol. 1. Translated by Hoban Wieland. Los Angeles: Semiotext(e), 2011.

———. *In the World Interior of Capital: For a Philosophical Theory of Globalization*. Translated by Wieland Hoban. Cambridge: Polity Press, 2013.

———. *Terror from the Air*. Translated by Amy Patton and Steve Corcoran. Cambridge, MA: MIT Press, 2009.

Sniderman, Adam, Mireille Broucke, and Gabriele D'Eleuterio. "Formation Control of Balloons: A Block Circulant Approach." In *Proceedings of the American Control Conference*, 1463–68. Chicago, IEEE, 2015.

Söllinger, Gunter. *S. A. Andrée: The Beginning of Polar Aviation, 1895–1897*. Moscow: Russian Academy of Sciences, 2005.

Spitzer, Leo. "Milieu and Ambience: An Essay in Historical Semantics." *International Phenomenological Society* 3, no. 2 (1942): 169–218.

Squire, Rachael. "Rock, water, air and fire: Foregrounding the elements in the Gibraltar-Spain dispute." *Environment and Planning D: Society and Space 34*, no. 3 (2016): 545–563.

Starosielski, Nicole. *The Undersea Network*. Durham, NC: Duke University Press, 2015.

Stewart, Kathleen. "Atmospheric Attunements." *Environment and Planning D: Society and Space* 29, no. 3 (2011): 445–53.

———. *Ordinary Affects*. Durham, NC: Duke University Press, 2007.

———. "Studying Unformed Objects: The Provocation of a Compositional Mode." *Fieldsights—Field Notes, Cultural Anthropology Online*, June 30, 2013. http:// culanth.org/fieldsights/350-studying-unformed-objects-the-provocation-of-a -compositional-mode. Last accessed September 11, 2017.

Stiegler, Bernard. *Technics and Time: The Fault of Epimetheus*. Vol. 1. Translated by Richard Beardsworth and George Collins. Stanford, CA: Stanford University Press, 1998.

Stone, Alison. "Irigaray's Ecological Phenomenology: Towards an Elemental Materialism." *Journal of the British Society for Phenomenology* 46, no. 2 (2015): 117–31.

Stone, Brad. "Inside Google's Secret Lab." *Bloomberg Business Week*, May 22, 2013.

Stone, Tessa. "Creating a (Gendered?) Military Identity: The Women's Auxiliary Air Force in Great Britain in the Second World War." *Women's History Review* 8, no. 4 (1999): 605–24.

Straughan, Elizabeth. "Touched by Water: The Body in Scuba Diving." *Emotion, Space, and Society* 5, no. 1 (2012): 19–26.

Stromberg, Elena. "'I Will Not Act before Understanding: Context Is Everything': The Work of Alfredo Jaar." *Los Angeles Review of Books*, December 8, 2013.

Stuart, Arthur. "A Nine-Acre Nest for Dirigibles." *Popular Science Monthly* (September 1929): 20–21.

Suzuki, Y., and K. Inoue. "Kamioka Underground Observatories." *European Physical Journal Plus* 127 (2012): 111–19.

Swedish Society for Anthropology and Geography. *The Andrée Diaries: Being the Diaries and Records of S. A. Andrée, Nils Strindberg, and Knut Fraenkel, Written during Their Balloon Expedition to the North Pole in 1897 and Discovered on White Island in 1930, Together with a Complete Record of the Expedition and Discovery*. Translated by E. Adams-Ray. London: Bodley Head, 1931.

Szerszynski, Bronislaw. "Reading and Writing the Weather: Climate Technics and the Moment of Responsibility." *Theory, Culture, and Society* 27, nos. 2–3 (2010): 9–30.

Tait, Peta. *Circus Bodies: Cultural Identities in Aerial Performance*. London: Routledge, 2005.

Teyssot, Georges. "Responsive Envelopes: The Fabric of Climatic Islands." *Appareil* 11 (2013): n.p.

Thibaud, Jean-Paul. "The Sensory Fabric of Urban Ambiances." *Senses and Society* 6, no. 2 (2011): 203–15.

Thomson, Jol, and Sasha Engelmann. "Intra-Acting with the IceCube Neutrino Observatory; Or How the Technosphere May Come to Matter." *Anthropocene Review* 4, no. 2 (2017): 81–91.

———. "TIANHE ≈ 天河: Parables of the Celestial River." *Aerocene Newspaper*, accompanying Tomás Saraceno's *Aerocene* installation at the Grand Palais, Paris, December 2015.

Thrift, Nigel. "Different Atmospheres: Of Sloterdijk, China, and Site." *Environment and Planning D: Society and Space* 27, no. 1 (2009): 119–38.

———. "Peter Sloterdijk and the Philosopher's Stone." In *Sloterdijk Now*, edited by Stuart Elden, 133–46. Cambridge: Polity Press, 2012.

———. "Understanding the Material Practices of Glamour." In *The Affect Theory Reader*, edited by Melissa Gregg and Gregory Seigworth, 289–309. Durham, NC: Duke University Press, Press, 2010.

Topham, Sean. *Blow Up: Inflatable Art, Architecture, and Design*. London: Prestel, 2002.

"Trailing Wires Cut Power Lines Serving Railways." *Independent* (St. Petersburg, Florida), September 18, 1940, 9. http://news.google.com/newspapers ?nid=950&dat=19400918&id=4-FPAAAAIBAJ&sjid=rFQDAAAAIBAJ&pg =4922,4387654. Last accessed May 26, 2014.

Tresch, John. "'The Potent Magic of Verisimilitude': Edgar Allan Poe within the Mechanical Age." *British Journal for the History of Science* 30 (1997): 275–90.

Truesdell, Clifford. "Jean-Baptiste-Marie Charles Meusnier de la Place (1754–1793): An Historical Note." *Meccanica* 31 (1996): 607–10.

Tucker, Jennifer. "Voyages of Discovery on Oceans of Air: Scientific Observation in an Age of 'Balloonacy.'" *Osiris* 11 (2006): 144–76.

Tully, John. *The Devil's Milk: A Social History of Rubber*. New York: Monthly Review Press, 2011.

Turner, Fred. "The Corporation and the Counterculture: Revisiting the Pepsi Pavilion and the Politics of Cold War Multimedia." *Velvet Light Trap* 73 (2014): 66–78.

Turner, Kate. "The Spectacle of Democracy in the Balloon Plays of the Revolutionary Period." *Forum for Modern Language Studies* 39, no. 3 (2003): 241–253.

Turpin, Etienne. "Aerosolar Infrastructure: Polities above and beyond Territory." In *Becoming Aerosolar*, by Tomás Saraceno, 169–83. Vienna: 21er Haus, 2015.

Twain, Mark. *Tom Sawyer Abroad, and Tom Sawyer, Detective*. London: Wordsworth, 2009.

Verne, Jules. *Around the World in Eighty Days, and Five Weeks in a Balloon*. London: Wordsworth Classics, 1994.

Virilio, Paul. *The Vision Machine*. Bloomington: Indiana University Press, 1994.

Volz, Wolfgang, and Peter Pachinke, eds. *Christo: Big Air Package*. Essen, Germany: Klartext, 2013.

Watt, Andrew. "The Exploration of the Upper Air." *Bulletin of the American Geographical Society* 42, no. 1 (2010): 37–51.

Webb, Stephen. *Measuring the Universe: Cosmological Distance Ladder*. New York: Springer, 1999.

Weidemann, Carolin. "Between Swarm, Network, and Multitude: Anonymous and the Infrastructures of the Commons." *Distinktion* 15, no. 3 (2014): 309–26.

Weizman, Eyal. *Hollow Land: Israel's Architecture of Occupation*. New York: Verso, 2012.

Werner, Alfred. *Introduction to the Graphic Works of Odilon Redon*. New York: Dover Publications, 2005.

Whitehead, Alfred North. *The Concept of Nature*. New York: Prometheus Books, 2004.

———. *Process and Reality*. Corrected ed. New York: Free Press, 1927.

Whitehead, Mark. *State, Science, and the Skies: Environmental Governmentality and the British Atmosphere*. Oxford: Wiley-Blackwell, 2009.

Williams, James. "Love in a Time of Events: Badiou, Deleuze and Whitehead on Chesil Beach," draft paper. https://whiteheadresearch.org/occasions/conferences/event-and-decision/papers/James%20Williams_Final%20Draft.pdf, last accessed September 7, 2017.

Willman, David. "How a $2.7 Billion Air Defense System Became a 'Zombie' Program." *Los Angeles Times*, September 24, 2015. http://graphics.latimes.com/missile-defense-jlens/. Last accessed November 11, 2015.

Winter, Caroline. "The Balloon's Big Moment." *Bloomberg Business Week*, September 5, 2012. http://www.businessweek.com/articles/2012–09–05/the-balloons-big-moment. Last accessed October 24, 2016.

Wong, Edward. "Beijing Issues Red Alert over Air Pollution for the First Time." *New York Times*, December 7, 2015.

Wylie, John. "An Essay on Ascending Glastonbury Tor." *Geoforum* 33, no. 4 (2002): 441–54.

———. "Landscape, Absence and the Geographies of Love." *Transactions of the Institute of British Geographers* 34, no. 3 (2009): 275–89.

———. "A Single Day's Walking: Narrating Self and Landscape on the Southwest Coast Path." *Transactions of the Institute of British Geographers* 30, no. 2 (2005): 234–47.

———. "The Spectral Geographies of W. G. Sebald." *Cultural Geographies* 14, no. 2 (2007): 171–88.

Xu, Qiaozhen, and Laurie Brown. "The Early History of Cosmic Ray Research." *American Journal of Physics* 55, no. 23 (1987): 23–32.

Ziegler, Charles. "Technology and the Process of Scientific Discovery: The Case of Cosmic Rays." *Technology and Culture* 30, no. 4 (1989): 939–63.

———. "Weapons Development in Context: The Case of the World War I Balloon Bomber." *Technology and Culture* 35, no. 4 (1994): 750–67.

Ziv, Amitai. "24/7 Big Brother: Gaza—The Technical Means That Observe the Gaza Strip 24 Hours a Day." *Marker*, August 1, 2014. http://www.rt.co.il/themarker. Last accessed October 16, 2016.

Ascension, allure of, 37; experience of, 38–40, 48–49, 60; luminosity and, 81; philosophy of, 13; as technical achievement, 57–58; transcendence and, 81–82; view from the ground and, 58; work of, 165

Ash, James, 5, 26, 75, 206

Atmosphere, 2–4; acts of balloon release and, 82, 88; allure of, 4, 6, 56–57, 76; bodies and, 8; as calculable/incalculable, 101–2; captivation and, 58; as concept, 6, 19; definition, 4; as emergent, 19; empiricism and, 7; entities and, 18, 26–27; envelopment and, 5–6, 28–32, 34, 42; as everyday idea, 19; as hard and soft, 147; immediacy and, 7–8; immersion in, 41, 212; materialism and, 6, 14, 28, 34, 105, 197; as meteorological, 4, 11, 20–21; the non-human and, 8, 11, 21; as object-target, 20; politics and, 8; promise of/problem with, 101–2; as relational, 19; remote sensing and, 50–51; scale and, 20; security and, 172; sounding and, 121–24; staging of, 10; as threshold, 141; as unformed, 12; variations within, 43, 52; volume and, 119; virtuality and, 27

Atmospheric address, 187–88, 192, 194, 254n9

Atmospheric experience, 10, 67, 70, 73, 116, 139

Atmospheric materialism, 6, 14, 28, 34, 105, 197

Atmospheric media, 11; GHOST and, 136–37, 187; infrastructure and, 204; elemental ontology of, 204; Project Echo and, 130, 136

Atmospheric publics, 110, 175, 194, 210

Atmospheric things, 195; allure of, 10–12, 75–77; balloon as a device for doing, 12, 14, 18, 22, 32, 34, 110, 197, 215; definition of, 10; elemental and, 83; envelopment and, 5, 28, 57; as excessive, 11; experiments with, 155; artistic/scientific/political operations and, 189–90, 194, 211; publics and, 175; Saraceno's work and, 211, 214; *Scattered Crowd* as, 118

AT&T, 124, 129

Attunement, 7–9, 148

Bachelard, Gaston, 38, 48

Baldwin, Thomas, 41

Balloon, allure and, 3, 73; as angel, 217; bombing and, 176–78; cultural history of, 12; as device for doing atmospheric things, 12, 14, 18, 22, 32, 34, 110, 197, 215; as disquieting, 49; drifting journeys of, 13–14; duration of, 107; as envelope, 30, 32, 43, 57, 67–68, 102–3, 160; as process of envelopment, 11, 30–31, 43; launch of as public spectacle, 58–59; levity and, 13; as lure for thinking, 18; as masculinist device, 91; as philosophical device, 12–13; relationship between horizontal and vertical movement of, 149; and sensing 42; as shape shifter, 18; superpressure, 133; as technical object, 102; as uncanny, 10;

Balloon, The (short story), 17–19, 22–23, 25–26, 32, 34, 166

Balloonatic, The (film), 161–62

Balloon drop, 1–2, 10

"Balloonmania," 61

Balloon prospect, 39, 41

Balloon seller, 61–62, 90

Banham, Reyner, 151

Bankhead, Tallulah, 155, 157

Banksy, 192

Barnes, Julian, 93–94, 98

Barrage balloons, 158–59, 168, 178–79

Beaton, Cecil, 153–57

Being-in-the-air, 33, 41–42, 47, 52, 60, 62, 146, 165, 167, 170

Bell Labs, 67, 70, 124–25

Bell Burnell, Jocelyn, 139

Media, 22, 59; atmospheric, 136–37; experiments with, 124, 139–40, 141–42; immersive experience and, 68–69
Media studies, 142–43
Mediation, 32, 59, 204
Melinex, 67
Membrane, 5, 29, 32, 67, 69, 105, 146, 172
Memory, 79–82, 88, 92–93, 98, 205
Memorialization, 86–88
Michelin, 64
Microwaves, 138
Milieu, atmospheric, 5, 9, 15, 29–30, 34–36, 48, 50, 84, 146–49, 215; elemental, 8, 10, 20, 32, 35, 102, 113, 124, 175, 196, 199, 203, 206–7, 212, 216; force of, 38; gaseous, 32; as immersive, 25; Pepsi Pavilion as, 68; vibrational, 143
Millais, John Everett, 107, 109
Millikan, Robert, 137, 176
Mirror Dome, 68
Mirror, The, 164
Mischer, Don, 1, 10
Montgolfier brothers, 31, 150
Monumentality/monumentalism, 111–14, 118, 160
Morton, Timothy, 25, 26, 32, 139
Motion, 27, 45, 202
Muesnier de La Place, Jean 149
Museo Aero Solar, 209–10
Mylar, 30, 65–67, 126, 131, 196

Nadar (Gaspard-Félix Tournachon), 41, 93
Nancy, Jean-Luc, 84
NASA, 124
NCAR, 131
New Zealand, 131, 201
Nieuwenhuis, Marjin, 8

Oceanic feelings, 47, 117
Object-oriented ontology (OOO), 23, 56
Ontology, 6–7, 14, 21–23, 25, 34, 56–57; elements and, 197, 205; envelopment and, 113; experiment and, 201; infrastructure and, 206; of media, 204; volume and, 104–5
Operation Outward (WWII), 179, 189
Operation Prospero (Cold War), 182
Operation Veto (Cold War), 182, 185, 187, 189–90
O'Sullivan, William 126–27, 131

Packer, Randall, 68
Pageos balloon, 130
Palestinian National Authority, 192
Parc André Citroën (Paris), 36
Parikka, Jussi, 136
Patents, balloon advertising and, 62–63, 65; project Loon and, 206
Pepsi Pavilion (Osaka Expo), 67–69, 73, 113, 127
Penzias, Arno, 138–42, 196
Perturbation, 7, 21, 138, 142, 147
Philippopoulos-Mihalopoulos, Andreas, 208, 213
Photography (aerial), 41
Pierce, John, 95, 125
Pilot/piloting, 4, 130, 148, 197, 209, 211
Pisters, Patricia, 192
Plastic bag, 210
Poe, Edgar-Allan, 61
Prayer, 89, 99–100
Presence, 32, 52, 57, 82–83, 107; absence and, 82, 86, 98; grief and, 85, 88; privileging of, 7–8, 26, 50; security and, 172–74, 190, 215
Prior, Jonathan, 142
Prisoner, The, 171–73, 189, 206
Project Gopher (Cold War), 186
Project Genetrix (Cold War), 186–87
Propaganda, balloons during Cold War and, 180–87; balloons during World War II and, 179–80
Prosthesis, 52
Psychoanalysis, 47
Psychological warfare, 181